Caledonia Dreaming

By John K.V. Eunson
and available from Hachette Scotland

Caledonication

Caledonia Dreaming

Caledonia Dreaming

JOHN K.V. EUNSON

First published in 2010 by
HACHETTE SCOTLAND
An imprint of
HACHETTE UK

1

Cataloguing in Publication Data is available from the British Library

ISBN 978 0 7553 1859 9

Designed by Viv Mullet
Typeset by Ellipsis Books Limited, Glasgow
Printed and bound in Great Britain by Clays Ltd, St Ives plc

Hachette Scotland's policy is to use papers that are natural, renewable and recyclable products and made from wood grown in sustainable forests. The logging and manufacturing processes are expected to conform to the environmental regulations of the country of origin.

HACHETTE SCOTLAND
An Hachette UK Company
338 Euston Road
London NW1 3BH

www.hachettescotland.co.uk
www.hachette.co.uk

Contents

Acknowledgements

I would like to thank Bob McDevitt, Rona Johnson and all those at Hachette Scotland and Headline Publishing Group for their support, encouragement, advice and above all tolerance in bringing *Caledonia Dreaming* to publication.

I would also like to thank my family and friends for their love and understanding in the process of bringing this idea from being merely an interesting suggestion to the book you are hopefully about to read. And in particular I would like to thank Aidan, Aidan, Anthony, Ashley, Belen, Carole, Conor, Duncan, Janne, Karen, Katy, Leila, Martin, Natalka,

Nicola, Sonia, Tristan and all the countless others not born in Scotland that make this country a far, far better place. *Al fin y al cabo somos hijos de Jock Tamson*, or in other words, we're all Jock Tamson's bairns after all. *Gracias amigos*.

Introduction

In those halcyon days of 2007, when you could buy your supermarket, super-strength bottle of vodka for £4.99 before ten in the morning, when nobody cared how grumpy Andy Murray was because wasn't it nice to have a half-decent tennis player after all these years and when the Royal Bank of Scotland's Sir Fred the Shred Goodwin was a master of the universe, I began to write my last book, *Caledonication*. I aimed to pen an entertaining history of Scotland from the Battle of Mons Graupius of AD84, through the tartan and flowing locks era of kings, queens, battles and clans and right up to the spiky, schizophrenic Scotland of 2007. A Scotland where a

Presbyterian son of the manse from Fife saw out the death throes of New Labour from his Downing Street bunker while a cheeky chappie from Linlithgow ran the devolved Government in Holyrood, preaching optimism and bonhomie, as the economy crumbled around them both. The furrowed brow versus 'Who ate all the pies?' – the centuries-old struggle for the soul of Scotland.

The careers of Gordon Brown and Alex Salmond also encapsulate another fundamental aspect of the history of Scotland – the Scot at home and the Scot abroad or, to put it another way, ever since James VI made his royal progress to London in 1603 to become James I of England, Scotland has been a nation where running your own affairs has always seemed a minor consideration compared to running somebody else's.

Many will be familiar with T. Anderson Cairns's famous prose *Wha's Like Us* that chronicles a host of items in everyday use, all invented or discovered by Scots. While researching for *Caledonication* I began to realise that the innovators listed by Anderson Cairns are only the tip of the iceberg (refrigeration, incidentally, was a Scottish invention) in terms of the globally renowned achievements originated by Scots in the fields of science, engineering, politics, education, health, culture, sport and much, much more.

I decided to find out more about those Scots who,

in most cases, left the kilts and the heather far behind them and begat the world we live in; not so much your Robert the Bruce, your Italian-born and bred Bonnie Prince Charlie or Oor Wullie, but more your Alexander Graham Bell who invented the telephone, John Logie Baird who pioneered television, the Scots who became UK prime minister, the Scots who were integral in the modern histories of, for example, Australia, Canada and the US and the numerous Scots who have been neglected at home but whose influence is still felt around the globe.

The criterion for selection was tough and completely subjective. There is a well-known question that you are probably aware of which goes, 'Can you name five famous Belgians?', to which the answer is of course Magritte, Hercule Poirot, Jean-Claude Van Damme, Audrey Hepburn and Tintin, with extra points for Adolphe Sax, inventor of the saxophone, and for either Kim Clijsters or Justine Henin; but even the good burghers of Brussels and Bruges might struggle to get far into double figures. You might ask a similar question, 'Can you name five famous Swiss who don't play tennis?' and you soon discover that Scotland is unusually blessed with internationally well-known figures.

To help narrow down the list to one hundred I decided to follow the traditional method of selection for the

Scottish football team for which you had to be actually born in Scotland to qualify; so, no qualification through long-lost Glaswegian grannies or five years' residency in the Lanarkshire commuter belt – which means that Spartak Moscow and Ireland footballer Aiden McGeady would be eligible for the list on the basis that he was born in Glasgow, but that the Scottish football team's most famous fan, Rod Stewart, would not be eligible as he was born in London.

Following this rule that you have to be, as far as my research can ascertain, born in Scotland has led to some notable and painful omissions – for his services to music and tall blondes, Rod the Mod would certainly have been a possible candidate for the final one hundred. Other luminaries who, despite their Scottish ancestry, must miss the list because of their place of birth include: Lord George Gordon Byron – the poet spent much of his childhood in Aberdeen with his Aberdeenshire-born mother but he was born in London; William Gladstone – the four times British prime minister was the son of a Leith-born merchant but this famous Liberal was born in Liverpool; Captain James Cook's father was a farm labourer from the Scottish Borders but Cook was born in Yorkshire and Ernest Rutherford, who split the atom, was born in Nelson, New Zealand although his father hailed from Perthshire.

Not on the list is a certain Joanne Rowling, born in Gloucestershire, who during her long search for a café in Edinburgh that served a decent skinny latte was able to write the international phenomenon that is *Harry Potter*. Further noteworthy omissions include two giants from the world of science, Joseph Black and William Thomson. The 18th-century physicist and chemist, Joseph Black – Scottish mother, Irish father but born in Bordeaux – while a professor at Edinburgh and Glasgow universities, isolated carbon dioxide and developed the theories of latent and specific heat. The 19th-century physicist, William Thomson, was born in Belfast but moved to Glasgow at the age of eight and, later, as a professor at the city's university found international fame for his work on thermodynamics. When Thomson was ennobled he took the title of Baron Kelvin of Largs after the river that flows past Glasgow University and the Ayrshire seaside resort and ice cream metropolis, and it is for his internationally recognised scale of absolute temperature to which he gave his name that he is remembered today; the Kelvin scale – and its more refined and exclusive variation, the Kelvinside.

It does seem harsh to exclude Lord Kelvin – resident in Scotland for some seventy years – and other luminaries who have lived in the country or can claim Scottish heritage but it does simplify the selection process, and for every exclusion, such as David 'call me Dave' Cameron – who was born in London but whose father hails from near Huntly in Aberdeenshire – there is an inclusion, such as the politician Cameron is most often said to resemble, Tony 'call me Tony' Blair, who, whether he likes it or not, was born in Edinburgh.

So, join me on this journey through the lives of one hundred Scots who have changed the world – and when I say, world, I don't mean Scotland and I don't mean Britain, I mean the whole, climate changing, waters rising, volcano erupting planet. Some of my select hundred never travelled south of Gretna, others left the land of their birth at the earliest opportunity; all have left an enduring legacy – although not always the one they might have hoped for.

Wha's like us? Damn few Belgians and even fewer Swiss who don't play tennis.

Iconic Scots

Who is Scotland's national icon? For many centuries Presbyterian Scotland frowned on the very idea of celebrity and idolatry – as well as frowning on a great many other things. Historical figures could be respected and, occasionally, appreciated but reverence was the domain of the Lord – and we don't mean Lord Kelvin. Whenever there has been a movement to select a national hero there has never been nationwide consensus as to who this figure should be. King Robert the Bruce, the victor at Bannockburn in 1314 and all that, might seem an obvious candidate, but in the days of post-1707 Union a figurehead such as Bruce, who had stood up for an independent Scottish nation, was not really in keeping with the ethos of the United Kingdom of Great Britain. The more romantic Bonnie Prince Charlie could be ruled out on similar grounds but also, and more problematically, because he was a misguided Italian whose disastrous foray into Scottish politics resulted in far more harm than good. And then there is Robert Burns [34], neither king nor prince, but a ploughman's son and Scotland's greatest poet, forgiven and even applauded for his licentious behaviour – but when all is said and done there is only so much haggis you can eat.

No, as far as icons are concerned, the Scots disapprove. Here, the icon is just one more example of an individual 'getting above themselves', a deeply suspect concept which engenders that wonderfully Scottish trait so that the more famous and important a person is, the more you look down upon them. The rest of the world, however, doesn't seem to share Scotland's unique view of inverse superiority, so here are a few potential icons to be getting on with . . .

OOR WULLLIE: AWBODY'S FAVOURITE FREEDOM FIGHTER

1 WILLIAM WALLACE (C1270–1305)

After all those centuries of pouring cold porridge on the very notion of a national icon the Scots only recently, and unexpectedly, discovered one in their midst who would inspire the entire world. Susan Boyle from Blackburn in West Lothian was forty-eight years old when she first 'dreamed a dream' on *Britain's Got Talent* in 2009, and within days more than one hundred million viewers worldwide had raised a collective, surprised and appreciative eyebrow in acknowledgement of her television debut.

However, as far as unlikely international sensations from Scotland that take a very long time to come about are concerned, SuBo's forty-eight-year wait must concede to the nearly 700 years which passed from William Wallace having his bowels fried in front of him in 1305 until the release of the Academy Award-winning blockbuster *Braveheart* in 1995.

At this point, we should also spare a thought for West Lothian-born film director, Michael Caton-Jones; 1995 was going to be his year. After finding critical acclaim at home and in the US, Caton-Jones had Hollywood backing to direct a big-budget, biographical epic in his native Scotland. With a star-laden cast headed up by Liam Neeson, stunning Highland locations and the life of Rob Roy MacGregor – one of Scotland's most popular historical figures – as the film's subject, everything was looking good. Unfortunately for Caton-Jones, you wait for decades for a Hollywood Scottish biopic and then two are released in the space of two months. Mel Gibson directed and starred as William Wallace in *Braveheart*: the film won five Oscars in the following year of 1996 including best picture and best director and, having grossed more than $200 million globally, is by far and away the most successful spaghetti Western ever set in Scotland.

From the very beginning, there was criticism among the praise for *Braveheart*, not least that most of the filming took place not in Scotland, but in the Irish counties of Meath, Wicklow and Kildare. Experts, critics and audiences alike pointed up the screenplay's many historical inaccuracies: not least the absence of any bridge in the Battle of Stirling Bridge, at which in 1297, Wallace achieved his only major victory over the English. Despite such inaccuracies, *Braveheart* proved as successful at the Scottish box office as it did elsewhere. Moreover, it has often been said that such an internationally popular film with such an openly Scottish nationalist and patriot perspective helped give the Scots a previously lacking self-confidence which helped towards the vote in favour of devolution in 1997; as Wallace said, 'You can never take away our right to tax-varying powers'.

The more you learn about William Wallace the more you understand why he has become such a popular icon both in Scotland and farther afield. Indeed, the more research you do, the more you realise that much of what is known about his life is based on rumour and hearsay and that Wallace the man is mostly Wallace the myth and can, therefore, become as much, or as little, of a hero and freedom-fighter as you want him to be.

Wallace was, possibly, born either near Paisley or in Ayrshire sometime in the early 1270s, apparently the second of three sons born to Sir Malcolm Wallace of Elderlie in Renfrewshire. Hardly anything is known about his early life, his education, or whether he was married or had children. The beginning of his resistance to the English occupation of Scotland has been claimed to have taken place in both Lanark and Dundee and hardly anything is known about his life between his defeat by the English in the Battle of Falkirk in 1298 and his capture by the English in 1305, except that he was believed to be in hiding in France, which in itself is quite an achievement when you are six foot seven tall. Wallace's reputation as a great Scottish patriot and martyr was founded predominantly on the epic poem by a Scottish minstrel called Blind Harry who collected and compiled events from Wallace's life in the oral tradition and, later, printed versions, *Blind Harry's Wallace*, would become one of the main sources for historians – although it has to be pointed out that, as his moniker suggests, Blind Harry was not an actual eye-witness to events and furthermore was born some 200 years after Wallace's death.

So, despite all the contradictions, inaccuracies and lack of historical evidence William Wallace has become Scotland's national icon, but not before one final piece of mythology slotted into place. In *Braveheart*, the scriptwriters were careful not to give the eponymous epithet to any one character but in the public's perception the heroic Wallace was, clearly, Braveheart. Yet if there has been one person in Scottish history that you associate with having a brave heart it is not Wallace, but the other hero of the Wars of Independence, King Robert I of Scotland or Bobby the Bruce himself. King since 1306 and vanquisher of the English at Bannockburn in 1314, when the Bruce was on his deathbed in 1329, having ensured that Scotland's independence had been finally accepted by both the English and the Church of Rome, he declared to his friend, James 'Black' Douglas, that to atone for all the sins of his life he wished that his heart be taken and buried in the Holy Land.

By 1330 Douglas and his compatriots, complete with embalmed heart in a casket, had arrived in Spain where they decided to join a Christian Spanish army which was attempting to overthrow the Moorish kingdom of Granada and, if they had time, they'd fit in a tour of the

Alhambra as well. Sadly, at the Battle of Teba in the Malaga region of Andalusia Douglas and most of the Scottish knights were killed. In his final, defiant moments – so the story goes – Douglas invoked the spirit of Bruce and uttered the legendary words, 'Onward braveheart, Douglas shall follow thee or die', then he threw the casket into the heart of the battlefield. Thankfully, the casket was retrieved from Andalusia and reburied in beautiful Melrose Abbey in the Borders, a fitting resting place for the true Braveheart of Scottish history.

The return of Bruce's heart has not, however, stood in the way of thousands of Scottish pilgrims determined to retrace the footsteps of Black Douglas by travelling annually to the Costa del Sol and fighting the locals.

AND ON THE SEVENTH DAY GOD RESTED AND GREW HIS BEARD

2 JOHN KNOX (C1512–72)

When you think about Scotland's best-known historic religious figures you might come up with a Holy Trinity of St Andrew, St Columba and St Margaret – but, of course, none of them was born in Scotland. St Andrew, the country's patron saint, was an Israeli/Palestinian whose bones, legend tells us – centuries after his crucifixion in Constantinople on a diagonal cross, hence the Scottish saltire – were brought to Scotland and buried in Fife. St Columba, founder of the monastery on the island of Iona and a leading figure in the conversion of Scotland to Christianity, was an Irish priest from a noble family exiled to Scotland. St Margaret, also known as Queen Margaret, the second wife of King Malcolm III and co-founder of the royal Canmore dynasty, known for her religious piety and anglicising the Scottish royal court, was an exiled English Saxon princess who may have been born in Hungary.

Scotland's most influential, home-grown, religious figure is, unquestionably, John Knox – the public face of the Scottish Reformation which during the 16th century transformed the country from Catholicism to Protestantism. Knox was instrumental, too, in the establishment of the Church of Scotland, or the Kirk, which over the ensuing centuries would export Scottish Presbyterianism – a version of Calvinism – around the world. For all his importance in history, Knox is a figure who Scots are reluctant to embrace. You will find little mention of him in the East Lothian town of Haddington where he was born except that the town's secondary school is named after him. He is buried in Edinburgh yet despite being the first Protestant minister to preach at the capital's St Giles' Cathedral his grave is said to be under the cathedral car park.

John Knox's influence on the Reformation in Scotland has been said by many to be over-emphasised; between 1546 and 1559, when Scotland's transformation away from Catholicism was beginning to take place, Knox was in exile in England, Germany and finally Switzerland. It was the time he spent in Geneva working alongside the theologian John Calvin that would persuade Knox that Calvinism was the ideal religion for the fun-loving Scots.

He returned home in 1559 and, with the Protestant Church of Scotland declared the country's official religion in 1560, was just in time to ensure that Scottish Protestantism would be radically Calvinist in nature; a far cry from the wishy-washy Anglicanism of the Protestant Church of England with its royal patronage, bishops, cucumber sandwiches and propensity for listening to *The Archers*.

While his enduring image may be that of a fire and brimstone demagogue sporting a wild beard and intent on bullying the demure young Mary, Queen of Scots [3], Knox was, in fact, a democrat, opposed to despots and a firm believer in both education and social welfare for all. This democracy did have its limits, however, as he loudly expressed in his tract against the influence of the fairer sex, *The First Blast of the Trumpet Against The Monstrous Regiment of Women*.

This form of Calvinism established in Scotland under

the influence of Knox (though again Knox's role has often been overstated) became known as Presbyterianism; power did not lie with the monarch or the government but with the presbyters, or elders, of the church who were answerable to nobody but God.

The word of God was the Bible and grace could be attained only through faith in Christ, no matter what Grace had to say about the matter. Hell would freeze over before the church that Knox built would bow to English bishops, but there were times during the 17th century when hell became a bit nippy as the Stewart/Stuart kings, now resident in England, tried to impose English Protestantism on their recalcitrant Scottish subjects. It was not until after the Glorious Revolution in 1688 finally deposed the Stuarts, and following the Act of Union in 1707, that the English finally accepted the independence of the Kirk, as long as they weren't too independent, in return for the Scots' acquiescence at their country being absorbed into the United Kingdom of Great Britain. God, it seemed, had become convinced of the merits of union, although was remaining neutral on the question of future devolved powers.

From the 17th century onward, wherever Scots went they took their cheery little brand of Presbyterianism with them. First stop was Northern Ireland where, in the 17th century, more than 100,000 Scots – mostly from Ayrshire, Galloway and the Borders – settled, making Protestantism the majority religion in the east of Ulster and Presbyterianism the predominant form of Protestantism in the province. When the island of Ireland was partitioned in 1921, the Presbyterian counties of Ulster stuck with king and country and, in return, gave the United Kingdom the Reverend Ian Paisley, James Nesbitt and the realisation that the words 'how', 'now', 'brown' and 'cow' could rhyme with 'pie'.

Next stop was North America. In the 18th century more than 200,000 Scots and Protestant Ulster Scots crossed the Atlantic and, although they were a relatively small minority of the Colonies' European immigrants, they played such an influential role in the American War of Independence that the exasperated British would often refer to the conflict as a 'Presbyterian' war – with once again, Protestant Scots, Protestant Ulster Scots and their descendants expressing their God-given right never, ever to be told what to do. Fifteen signatories to the Declaration of Independence were Presbyterians and of the forty-three US presidents to date, ten have come from Presbyterian backgrounds, although, as far as we know, President Obama is not yet a subscriber to the Church of Scotland's monthly magazine *Life and Work* – the perfect Calvinist title, for what, after all, is one without the other?

Today, there are tens of millions of Protestants around the world that adhere to the Presbyterian belief that black never goes out of fashion. However, being Presbyterians and somewhat opinionated, there have been many disagreements, divisions, schisms and splits over the centuries, which makes agreeing on a total worldwide figure somewhat difficult, although heaven only knows what exactly is the difference between, for example, the Free Presbyterian Church, the Free Church and the United Free Church, and even God gets a migraine when he has to think about it too much. It has long been said that a Presbyterian could start an argument in an empty church and in Scotland empty kirks have become more and more prevalent, or more sacrilegiously converted into flats and even theme pubs. One suspects that John Knox, the founding father of Presbyterianism, has in recent years probably done much turning in his grave beneath St Giles' car park – except on the Sabbath, of course.

WHAT DO YOU GIVE THE WOMAN WHO HAS LOST EVERYTHING?

3 MARY, QUEEN OF SCOTS (1542-87)

A tall and striking woman whose beauty inspired the love and devotion of many and the exasperation of many more, she was married at an early age as a matter of political expediency. She produced the prerequisite heir to the throne, and could have become queen of Britain and the symbol of a new golden age, but she was undone by being too controversial for the establishment of the time and her truly terrible choice of men. Her life would ultimately be that of a tragic heroine and she would die in a sudden and violent manner, and would have the mixed blessing of being remembered around the world long after her death. Yes other than hair colour, there really is not that much difference between the lives of Diana, Princess of Wales and Mary, Queen of Scots. It just goes to show that in history if you are a woman in the wrong place at the wrong time, they will always get you in the end.

Scottish history does not really do women. While the English have seemed always to find comfort and solace when they have a strong woman as their figurehead or leader – Boudicca, Elizabeth I and II, Victoria, Margaret Thatcher, Anne Robinson etc. Scotland has no equivalents. Our famous monarchs, soldiers, scientists, politicians, inventors and sport stars have been almost universally male, with only an occasional woman such as Jacobite heroine Flora Macdonald being allowed in for some token gender balance. Good old Flora, didn't she do well? She went on a boat, once. And Scotland does not even seem to follow the maxim, 'Behind every great man there is a great woman'; we rarely mention the spouses of our most famous sons, more, 'Behind every great man there is a woman making his tea'. The brief reign of Mary, Queen of Scots is the only time in history when a Scottish woman has ruled Scotland, and although that was more than 450 years ago it appears to have traumatised the nation so much that the thought of Nicola Sturgeon ever becoming First Minister makes some grown men, and for that matter some grown women, go a little twitchy.

Mary Stuart was born in Linlithgow in 1542. People born in Linlithgow, including First Minister Alex Salmond, are known as 'black bitches' in recognition of the town's symbol of a faithful dog who brought food to his imprisoned master and for doing so, the dog was tied to an oak tree on an island. During her lifetime Mary was called a black bitch, and much worse besides. She was the only surviving child of James V and the French Mary of Guise and became queen when she was less than a week old, which even by Scotland's standards of having young monarchs was an unusual state of affairs and with baby's presents consisting of the Scottish Crown Jewels rather than your standard rattle.

Immediately the infant queen became a vital pawn in the never-ending battle for dominance between England and France. She was monarch of Scotland and in 1548, at the age of six, was sent to France where she was promised in marriage to François, the four-year-old heir to the French throne, much to the annoyance of the English who for the past six years had been trying to persuade the Scots that Mary should marry the future Edward VI, the only son of Henry VIII, so uniting the two crowns. Ten years later, Mary and François were married and François succeeded to the throne in 1559 as Francis II.

The marriage was the culmination of the Auld Alliance with those two great nations, France and Scotland, at long last united politically and dynastically – a meeting of minds and cultures which transformed le coq au vin into le coq au leekie and combined the French kiss with its Glasgow equivalent. It had taken 250 years for this union to come about but sixteen months later it was all over. In 1560 the sixteen-year-old Francis II of France died and his teenage Catholic widow, who had been living in France for 12 years, was packed off back to a Scotland that had inconveniently turned Protestant in her absence.

Eight years later, Mary was on her travels again, fleeing for her life to England, now an ex-Queen of Scotland as well as France, with a second husband murdered and a third on the run somewhere in Scandinavia. She had been forced to abandon her crown in 1567 and had been forced to abandon her infant son, James VI, who had been proclaimed king upon her abdication. Mary turned to Elizabeth I for sanctuary and the relationship between the two women – who were first cousins as well as fellow monarchs – has always remained a fascination for writers and historians.

Mary may have lost two crowns, but she still had ambitions to go for the hat trick. Elizabeth of England was the last of the Tudors when she became queen in 1558 and showed little inclination to marry, let alone produce a brood of little redhead heirs. Thus, it was Mary, the queen whom nobody wanted, who was next in line to the throne through the marriage in 1509 of James IV of Scotland, Mary's grandfather, and Margaret Tudor, daughter of Henry VII and sister of Henry VIII. This union was known as the marriage of the Thistle and the Rose – an appropriate description for the prickly relationship between the two countries.

Furthermore, as far as Catholic Europe was concerned Protestant Elizabeth was not a legitimate monarch as her mother, Anne Boleyn, had only become queen after Henry divorced his first wife, Catherine of Aragon, against the express wishes of the Catholic church. And so Mary, as the next in line to the English throne and also a Catholic, returned to her role of political football – a cause without a rebellion.

For nineteen years Elizabeth kept Mary prisoner, refusing to meet her younger and more glamorous cousin in person and constantly having to weigh up whether it was more trouble to keep her alive or have her done away with, before finally deciding on the latter course in 1587. Mary was executed at the age of forty-four at Fotheringay Castle in Northamptonshire, her status as a tragic heroine and now martyr enshrined forever. One year later in 1588, Phillip of Spain sent his infamous Armada to England to depose Elizabeth from the English throne, with the execution of the Catholic queen proving one of the catalysts for his attempted invasion, but, even dead, anything even remotely connected to Mary was bound to end in disaster.

There are numerous books, poems and songs about Scotland's most famous woman; there was even a Hollywood film, *Mary of Scotland*, made in 1936 and starring Katharine Hepburn as Mary – appropriate casting with Hepburn claiming Scottish descent from one James Hepburn, Earl of Bothwell, who was Mary's third husband.

Despite the hardships Mary endured in life, when Elizabeth I finally died in 1603, it was Mary's son, James VI, who succeeded her as king of England and Scotland and it was not the Tudor but the Stuart/Stewart line – so good they named them twice – which would prevail and, thus, it is from Mary that today's royal family are descended. In 1612 James buried his mother in a magnificent tomb at Westminster Abbey. While Mary's tomb lies next to that of Elizabeth, the two are, tellingly, out of each other's sight – in death as they were in life. Elton John was asked to write a special song for the occasion, but it wasn't very good.

OF ALL THE GIN JOINTS IN ALL THE WORLD, I WALKED INTO THE ONE WITH IRN-BRU AS A MIXER

4 ALEXANDER GORDON (18TH CENTURY)

Whisky, whisky, whisky. If you were visiting Scotland for the first time you might expect the Scots to drink nothing else and you would, therefore, be somewhat bemused to discover that they drink everything else but whisky, as Scotland's Presbyterian history notwithstanding, its countrymen have an exceedingly catholic appreciation where alcohol is concerned. This willingness to experiment continued when the Scots travelled abroad, where they embraced and enjoyed new alcoholic delights, from claret in France to rum in the Caribbean, with the only known exception being raki, which no combination of mixers has ever made palatable and also partly explains why so few Scots ever settle in Turkey.

Once the Scots had discovered a new tipple and found it to be good they were keen to spread the word and, ideally, make some money out of it. For example, it was Scotsmen who founded two of the world's most famous brands of port, Sandeman and Cockburn's. Perthshire-born George Sandeman began importing port and sherry from Oporto in Portugal to his offices in London in 1790 and in 1815, Robert Cockburn came to Portugal's Douro Valley from where he imported port and fine wines to the port of Leith. Cockburn's Special Reserve became the drink of choice for after dinner relaxation, although social disgrace was a clear and present danger if you neglected to pass the port in the requisite fashion.

Throughout the 1970s and 80s Cockburn's winter television advertising campaign became a seasonal staple, firmly establishing the tradition which, after demolishing your turkey and Christmas pudding you were expected to consume port and chocolates while watching the *Only Fools and Horses* Christmas special. Invariably, as most viewers would then proceed to fall asleep, the port would end up all over their brand new Christmas jumpers, which considering the jumper in question was either stripy or beige and was being worn only under sufferance this mishap was, as the Cockburn's slogan promised, 'Well Worth the Wait'.

Even more popular than port however, was a gin from the stills of a company by the name of Gordon's, founded in London by a Scotsman, Alexander Gordon, in 1769. Gin was very widely drunk in England during the 18th century as it was easy to produce and inexpensive, but as much of it came from poor-grade grain the drink's quality was often questionable, although its effect was potent – the term 'gin-soaked' comes from this time.

Alexander Gordon was determined to come up with a more palatable gin and after much experimentation at his distillery in Clerkenwell came up with a London dry gin which had been triple-distilled and included juniper berries, coriander, angelica and one other botanical. So pleased was Gordon with his recipe that the final botanical was not made public and would, instead, be handed down from generation to generation in a ceremony so secret that no more than twelve people in the world would ever know the complete recipe at any one time – and any divulgence of the secret would be punishable by death or, worse, being forced to drink Malibu.

Gin found a new market in the 19th century when it began to be drunk mixed with tonic water. This cocktail came about by accident in India where, as a means of combating malaria, the British had been encouraged to drink a tonic water containing quinine. On its own, the tonic water tasted extremely bitter but when added to gin the flavours were found to complement each other and the Brits were soon drinking so much of the stuff that any bloodthirsty mosquito risked severe alcohol poisoning.

Gordon's distinctive green glass bottle became ubiquitous not just in Britain but, with thanks to the enthusiastic support of the British Navy, around the globe and in 1925 Gordon's gin, the world's best-known brand, received a royal warrant. The tipple met with approval among members of the current Royal Family, with the Queen Mother in particular said to partake of the occasional bottle or three – which considering she lived to be 102 did wonders for the brand's reputation. A further accolade came Gordon's way when the gin played its part alongside Humphrey Bogart and Katharine Hepburn in the 1951 classic *The African Queen* – when Bogart took the award for best actor, his Oscar had been especially made out of green glass.

Today, the multinational drinks company Diageo owns Gordon's Gin; it sells more than one hundred million bottles a year that makes it the best-selling exported gin in the world. Nearly seventy-five percent of gin is drunk with tonic and the image of that well-deserved G & T after a hard day at the office has been carefully cultivated with slogans such as, It's Got to be Gordon's: this tagline would be quietly dropped at the 2010 General Election, which was somewhat ironic as the person in most need of a Gordon's was a certain Mr Brown.

HERE COMES THE RAIN AGAIN

5 CHARLES MACINTOSH (1766–1843)

In the early 19th century John McIntosh, an American-born son of an immigrant from Inverness, settled in Ontario, Canada, and planted some wild apple trees in his garden. One of those trees produced a crop of red apples and they were given the name McIntosh Red that subsequently became one of the most popular apples in Canada. Many years later when those clever, bespectacled people at Apple were looking for an apple-related name for their new home computer project, Cox's, Granny Smith and Golden Delicious were all considered before they plumped for the McIntosh – which was then changed to Macintosh to avoid any confusion at the greengrocer's. In 1984 the Apple Macintosh was launched and the Mac became one of the computer world's top success stories. But long before the days of personal computers there was another Mac that no businessman could do without.

Glasgow-born Charles Macintosh worked as a

chemist in his home city where, during the 1790s, he collaborated with Charles Tennant – born in Ochiltree, Ayrshire – on the invention of bleach liquor and, later, on bleaching powder, made using a mix of chlorine on slaked lime. It was Tennant who took out the patents, for the liquor (1798) and for the more user-friendly powder in 1799, and it was he who would make his fortune after setting up a chemical works, Charles Tennant & Co., at St Rollox, Glasgow in 1800 in order to manufacture said bleaching powder. The product found a huge market at home and abroad with the St Rollox site, becoming the largest chemical works in the world, as well as the then-largest air polluter in the world. The Tennant family ignored the smog and joined the British aristocracy, but Macintosh, too, would have his moment in the sun, or to be more accurate, his moment in the rain.

Macintosh set up his own textile factory in Glasgow, but continued his chemical experimentations and, following a waste-not-want-not policy, was determined to find a use for naphtha, a recently discovered by-product of tar. In 1818 an Edinburgh doctor, James Syme, had discovered the waterproof properties of rubber dissolved in naphtha but had not progressed his finding. Macintosh came up with the idea of sandwiching the naphtha-dissolved rubber between two pieces of cotton and found that the resultant material was waterproof.

Macintosh patented his discovery in 1823 and, the following year, began manufacturing waterproof coats in his textile factory. The first Macintosh coats were rather stiff, prone to smell and inclined to melt in hot weather or crumble in cold, but the inevitability of rain on the west of Scotland rendered these risks worth taking. Further innovations in the rubberising process improved the coats' wearability and sales increased. Confusingly, the item was named the Mackintosh, with a 'k' added to Macintosh for no good reason, and over the years this would be shortened to mac or mack.

Breathable fabrics and plastics would eventually replace rubber, but the name Mackintosh remains a generic term for raincoats and trench coats – in honour of its inventor. Throughout the world and over the decades, whenever and wherever there was a chance of rain, a Mackintosh in fetching grey was an item that no household wanted to be without. In the 21st century the mac has experienced a renaissance, after years of consignment to the fashion wilderness. In vogue or not, it certainly could not be faulted for durability as proven by Peter Falk in his role as Lt Columbo; for thirty-five years he wore the same mac in every episode, despite there being absolutely no sign of rain in all that time.

THE COOS DE GRASS

6 HUGH WATSON (1780–1865)

There was a revolution in the late-18th century, often neglected compared to the Industrial, French and American ones: the Agricultural Revolution began in Britain and spread throughout Europe and included enclosing land and crop rotation, and it was a Scot who would come up with the key invention fundamental in transforming the countryside.

Up until the 1780s a good threshing was what upper-class Scots would call being disciplined at public school, but around 1786 Scottish miller Andrew Meikle from East Lothian developed a threshing machine which could separate the grain of a crop from its stalks and husks, a job that previously could be done only by hand. The original threshing machines still had to have the grain inserted by hand and were pulled by horses, but as improvements were made to Meikle's machine into the 19th century, more and more farm labourers and horses became unemployed and had to move to towns and cities to find work in the factories and mills, although the horses in particular found working the looms quite tricky.

The 19th century was a difficult period for horses but one breed that prospered as a beast of burden was the Clydesdale, first bred in the west of Scotland in the 18th century, and this strong, hard worker became popular throughout Britain then, later, in the US and Australia. The Shetland pony, on the other hand, is an ancient breed and has been resident for thousands of years on the islands it is named after. Combining strength and resilience with lack of inches the ponies were first used outside Shetland in the 19th century as pit ponies. In the mines they would rarely see daylight, which although cruel would not have been too dissimilar to spending their winters in Shetland. Its diminutive size and sweet, cuddly façade belie a potentially headstrong, stubborn animal that can turn on you in an instant and head-butt you somewhere around the knee area – not so much the 'wee hard man', but the 'wee hard pony'. Whether it was through the influence of the Shetland pony or the many Shetlanders who left the island archipelago, the word Shetland became so associated with all things small and hairy that when a miniature rough collie with a double coat was bred in the early 19th century it was named the Shetland sheepdog, despite having no connection with the islands.

Talking of dogs, the Skye terrier is but one of four breeds from the Highlands and Islands that have found favour around the world. Terriers in Scotland date back to at least the 15th century when they were bred as hunting dogs and were eventually given the generic name of Skye terriers; indeed, no journey on Skye was complete without having a small dog biting your ankles. At the end of the 19th century the Skye terrier was divided into four distinct breeds: the Skye, the Scottish or Scottie, the Cairn and the West Highland White or Westie.

All of the four breeds would become popular throughout the English-speaking world and beyond. In the mid 20th century the Westie became a must-have Hollywood accessory and when in 1961 Walt Disney released the film *Greyfriars Bobby: The True Story of a Dog*, it was a Westie that controversially played the title character, even though in reality Bobby was actually a Skye terrier, which slightly contradicted the title of the movie.

It was a Cairn terrier called Terry that played Toto in the 1939 film *The Wizard of Oz*, accompanying Judy Garland as Dorothy on her adventures from Kansas, but of all the Scottish terrier breeds it is the Scottie that has arguably become the most prestigious. Two US presidents have given them houseroom: Franklin D. Roosevelt owned Fala, named after the small village in the Scottish Borders, and his memorial statue in Washington features his faithful companion by his side; more recently the White House was home to Barney, George W. Bush's favourite pet throughout his presidency, although despite its Scottish hunting origins it never could find those pesky weapons of mass destruction.

Despite the traditional iconic status bestowed upon the Highland cow, there are two other Scottish breeds of cattle which have proved more successful both as exports and in terms of the basic cow skills of producing milk and beef – these are, respectively, the Ayrshire and the Aberdeen Angus.

Ayrshire dairy cattle – with their distinctive red and white hides – originated in the 18th century in the region's farming districts of Cunninghame and Dunlop. The breed was first exported to North America around the 1830s and their hardiness, longevity and high-quality milk led them to become one of the predominant dairy breeds there, leading to the famous John Wayne misquote, 'Get off your horse and drink your Ayrshire milk'.

While south-west Scotland breeds dairy, the current leading beef stock in North America and Australia hails from the north-east. In 1808 a young farmer from Angus called Hugh Watson acquired a herd of six cows and one bull and set out to make it the finest in the country. He crossed his stock with the best of the local black hornless cattle in Aberdeenshire and Angus, plus a few ringers from other parts of Scotland and continued to build up the herd this way over the ensuing fifty years.

Watson was not the only farmer to successfully cross Aberdeenshire and Angus beasts, but he is seen as the founder of the Aberdeen Angus breed we know today. The results he achieved were mainly due to the sterling efforts of two of his herd, a bull and a cow, who between them either sired or bore so many calves that the majority of all the Aberdeen Angus cattle in the world today are said to be descended from them. The bull went by the name of Old Jock, the cow was called Old Granny – both of which we can assume were not the names they were first given.

The first Aberdeen Angus were exported to Kansas in 1873 and the breed soon found a firm and enduring foothold with Stateside farmers impressed by their high-quality beef, their ability to prosper in different climates and their ability to crossbreed successfully – something which would undoubtedly have made Old Jock proud. In the United States, the Aberdeen Angus remains one of the most popular beef cattle breeds today, although in the US it is called the Angus rather than the Aberdeen Angus, due to Americans' inability to come to grips with the Aberdonian accent.

I WISH IT COULD BE CHRISTMAS EVERY DAY

7 DAVID DOUGLAS (1798–1834)

If Scotland can lay claim to being the spiritual home for celebrating Hogmanay with Robert Burns's [34] *Auld Lang Syne* as its theme tune, the same claim cannot be made for the festival of Christmas. For centuries Presbyterian Scotland celebrated the birth of our Lord by ignoring the whole thing: no fuss, no extravagance, no Slade and even going to the Kirk would be scheduled around work commitments. It was not until 1958 that Christmas Day became an official public holiday in Scotland and if you want to enjoy a traditional Scottish Christmas then going to your place of employment for eight hours would be a good way to start, although if your computer is a little slow don't expect much in the way of IT support.

However, Christmas cannot escape its Scottish connection and thus, dating back to the 19th century and without which no festive season would be complete, is the Christmas tree. The idea for a festive tree originated hundreds of years ago in Germany and was brought to Britain by the German Prince Albert and to America by German immigrants. Traditionally, in Britain the Scots pine has been the Christmas tree of choice, but Americans favour the Douglas-fir and the Fraser fir – both named after Scottish botanists.

The Fraser fir comes from the Appalachian mountains and was named after John Fraser who was born near Loch Ness, while the Douglas-fir comes from Western America and is named after David Douglas from Scone, the ancient capital of Scotland and the original and true home of the Stone of Destiny, in Perthshire. In 1824 Douglas set out on an expedition to North West America in search of new trees and plants and returned to Scotland with a collection of more than 200 – including the tree which now bears his name – that he duly introduced to British soil.

Technically, the Douglas-fir is not a fir, but an evergreen tree and, confusingly, its Latin name is most commonly given as *Pseudotsuga menziesii* in honour of a rival Scottish botanist from Perthshire, Archibald Menzies, who first discovered the tree on Vancouver Island some thirty years earlier.

The Douglas-fir has been successfully cultivated in many countries and in America is the tree most commonly grown for timber. The second tallest-growing trees in the world, these conifers can grow to an impressive 400 feet – ideal for very tall houses and for the Douglas squirrel, also named after David Douglas. The grey Douglas squirrel is you will be glad to hear a nice squirrel, nothing like the nasty Easter Grey squirrel from America that came over to Britain, stealing all our local squirrels' jobs and homes.

Douglas would make one further expedition, this time to the Pacific Sandwich Islands in 1833, before he met a grim end the following year when he fell into a cattle trap and was gored to death by a wild bull, although how such an experienced explorer as Douglas missed the wild-bull-sized shape in the middle of the ground has never been fully explained.

KILMARNOCK NO MORE

8 JOHNNIE WALKER (1805–57)

First of all, the whisky basics: the name whisky is an English mispronunciation of the Scottish Gaelic *uisge beatha*, 'the water of life'; malt whisky is made by distilling malted barley and water, grain whisky is distilled from malted barley plus some other grain and water, while blended whisky is a blend of malt whisky and grain whisky; whisky is often called Scotch in honour of the country it comes from; legally, Scotch whisky must be made in Scotland and must also be allowed to mature in Scotland in oak casks for a minimum of three years before it can be called whisky; there are five main regions of whisky production in Scotland – Campbeltown on the south of the Kintyre peninsula, the island of Islay, the Lowlands, the Highlands (including the other islands) and Speyside, which has more distilleries than any other.

Some people assert that it was the colonising Irish who introduced whisky to Scotland sometime in the 5th century, but it appears that for as long as there has been barley and water there has been some form of whisky of varying proofs and smoothness. The inhabitants of Scottish monasteries were especially keen on distilling the 'water of life', which was said to have medicinal qualities and, also, made asceticism a little more bearable. The first written mention of whisky in its Latin form *aquae vitae* dates back to June 1494 and a reference to the amber nectar in the Exchequer Rolls of Scotland is a commission from James IV for, 'Eight bolls of malt to Friar John Cor' – a boll being a large measure of grain and eight bolls of malt being sufficient to make 1,500 bottles of whisky. Friar John Cor resided in Lindores Abbey in Fife where in those pre-Reformation days self-denial was clearly a thirsty business.

It was not long before the authorities moved from enjoying whisky to seeing its popularity as a means to make money and the first taxes on production began in 1644. The taxes would become increasingly harsh over the ensuing 150 years and resulted in most whisky being produced in illicit stills. The British whisky industry did not begin to grow until 1823 when Parliament passed the Excise Act, which relaxed the restrictions on whisky distilleries, and this, combined with improvements in production and the adoption of a continuous distilling still first invented by a Scottish distiller called Robert Stein, began to boost the industry.

At first, whisky-drinking was confined to the Scots alone, but when in the 1880s the American phylloxera bug was introduced accidentally into the vineyards of France – possibly by a man striding out in a red jacket, who might have had a Scottish accent – it decimated wine and brandy production. Imbibers in England and Europe thus found themselves looking for a new digestif with which to enjoy their after-dinner cigar, sampled this obscure northern spirit and found it to be good. Then, in 1920 when the US Government brought in prohibition and banned the sale and production of alcohol – which as far as Southern Comfort and root beer were concerned was perfectly legitimate – Scottish whisky producers saw it as their patriotic duty to the many millions of Americans of Scottish descent to ship as much whisky as humanly possible to Canada and surrounding countries from where it could be illegally smuggled across the US border. By the time prohibition was finally repealed in 1933 Scottish whisky had been established as the favourite tipple for millions of Americans – to whom it was known simply as 'Scotch' – and, now that it was a legal import, Scottish producers no longer had to make donations to Italian-American 'opera lovers'.

The traditional whisky image tends toward the vision of a misty, Highland loch or roaring log fire and a large, single malt. But malt whisky makes up only fifteen percent of whisky exports and it is, in fact, the blends which sell abroad – and the three best-selling blends on the export market are Chivas Regal, Ballantine's and the former pride of Ayrshire, Johnnie Walker.

Chivas was founded in 1801 and is produced at the Strathisla Distillery in Speyside from where it sells more than four million cases a year. The founder of the blend, James Chivas, ran a grocer shop in Aberdeen and it was he and his brother, John, who set about producing a high quality blended whisky. In 1843 Chivas Brothers received a royal warrant to supply Balmoral Castle with provisions and it was not long before Victoria and Albert had become partial to a wee Chivas or two to make the local rowies more palatable – and then they told all their royal chums about how good the whisky was. Chivas Regal was launched in 1909 and became an international bestseller, renowned for its smoothness, and was a firm favourite with Frank Sinatra, Dean Martin and the rest of the Rat Pack.

Ballantine's is produced in Dumbarton. The brand is named for the Ballantine family from Edinburgh; in 1827

George Ballantine began to sell whisky in the family shop and by the mid-19th century the family were developing the blend that now sells more than five million cases a year. Ballantine's Finest is the top seller in Europe, a not inconsiderable achievement when one considers that thanks to the huge growth of Scotch whisky in Europe over the past thirty years, France and Spain both now match the US as the biggest whisky markets in the world.

Impressive as they are, sales of all the other Scottish blends fall far behind the sixteen million cases a year shifted by the market leader, Johnnie Walker. The story of the world's most popular whisky begins in 1820 when John Walker took over the family's grocer shop and began to sell his own brand, Walker's Kilmarnock Whisky. Walker was only fifteen years old at the time – the average drinking age in Ayrshire today – and when he died in 1857 the shop was taken over by his son, Alexander Walker.

Alexander turned the business into a wholesaler's and, ten years after his father's death, he developed a new blend; he called it Old Highland Whisky and sold it in square bottles. In 1908, in honour of their grandfather, Alexander's sons changed the name from Old Highland to Johnnie Walker and introduced the different coloured labels – red, black and a short-lived white label. They also came up with the idea of the trademark striding man who, red-jacketed and complete with top hat and cane, was based on their grandfather and became the brand's worldwide logo.

Today, Ballantine's and Chivas Regal are both owned by the French firm Pernod Ricard, while Johnnie Walker belongs to the drinks giant Diageo. When Diageo decided that the Johnnie Walker brand needed a 21st-century relaunch, a revamped striding man became the focus of the marketing campaign, although Diageo neglected to mention that rather than walking purposely toward the pub or home for a wee dram, in 2010 the striding man would be walking purposely out of Kilmarnock never to return.

In 2008, Scotch whisky exports achieved more than £3 billion worth of sales for the first time. Worldwide demand has never been greater and the continuing success of the magical mix of barley and water lifts Scottish hearts and makes the nation proud. So, in the words of the great Gaelic toast to kith and kin, 'Slàinte mhath, and mine's a bacardi breezer' – we may make oceans of the stuff, but you will never catch us actually drinking it.

THE FIRST BOY SCOT

9 WILLIAM ALEXANDER SMITH (1854–1914)

The honour of being the world's oldest uniformed youth organisation does not belong to Britain's Boy Scouts –

founded by Robert Baden-Powell in 1907 – nor does it belong to the Boys Scouts of America, but it is held by Scotland's very own Boys' Brigade, which was formed in Glasgow in 1883. The founder of the BB was William Alexander Smith, who was born in Thurso and became a businessman in Glasgow. In his spare time Smith was both a Lieutenant with the reserve unit of the Lanarkshire Rifle Volunteers (LRV) and a Sunday school teacher for the Presbyterian Free Church in Glasgow. Smith became aware that while the teenagers in the LRV were well-behaved and enthusiastic, the older boys at Sunday school, only a few years their junior, were distracted and bored at the prospect of Bible class with even the lessons concerning Adam and Eve being naked failing to keep their attention.

Smith came up with the idea of combining the religious with the military and, with that in mind, he set up his first Boys' Brigade company complete with military-style uniform and army-style discipline and leadership structure. He had his youngsters drill like soldiers and partake in physical activities and this he interspersed with prayer and religious instruction. These were the days before supermarket lager deals and the Boys' Brigade immediately attracted a keen membership. Its appeal grew even stronger when in 1886 Smith introduced his first Brigade summer camp in Argyll;

twelve months of marching, squat thrusts and endless uniform cleaning would be offset by a trip to the countryside and a week of camping in the rain-soaked, midge-infested, nettle-stinging middle of nowhere. The archetypal Scottish holiday where all potential for fun and frolics are thwarted at every turn; because as every Calvinist knows it is best to come to terms with disappointment when you're young, because that's how life is.

William Alexander Smith gave the Boys' Brigade its symbol of the anchor flanked either side with the letter B and he came up with this mission statement:

> The advancement of Christ's Kingdom amongst Boys and the promotion of the habits of obedience, reverence, self-respect and all that tends towards a true Christian manliness.

Despite the statement's complete absence of catchiness, the Boys' Brigade continued to gain members throughout Britain and her empire, with hundreds of companies and battalions totalling more than 100,000 boys established by the late 19th century. In 1907 Smith asked Baden-Powell to inspect 7,000 Brigade members in Glasgow. While he was very impressed by their discipline and organisation, Baden-Powell found the ethos a little too focused on military drilling and religious instruction and offered to expand the boy's activities by introducing them to the scouting techniques he had excelled at in the Boer War.

As an experiment, Baden-Powell selected a company of boys from various English brigades and, also, from different social backgrounds and took them to a camp in Dorset. The boys had a great time lighting fires, whittling sticks and stalking rabbits and despite their differences in upbringing all united as a group to pick on the fat boy. And that so impressed Baden-Powell that he wrote up the boys' programme of activities for his best-selling *Scouting for Boys* published in 1908.

Baden-Powell had built up a friendship with William Alexander Smith and had not intended scouting to rival the BB, but boys around Britain and, before long, around the world were keen to embrace scouting and the movement was begun. Initially, Smith and the Boys' Brigade were non-plussed by this upstart competitor, but the Scouts' membership soon overtook that of the BB and the scouts' khaki uniform and jaunty neckerchief neatly fastened with the obligatory woggle began to supplant the Presbyterian's blue togs and plain white haversack.

However, the Boys' Brigade did maintain steady growth throughout the 20th century and today the organisation has a worldwide membership of 500,000 youngsters unwilling to dib-dib-dib and secure in the knowledge that, despite being outnumbered fifty-to-one, in a straight fight between the two organisations not one woggle would still be in place. Whatever a woggle is.

WHEN LABOUR REALLY WAS NEW

10 KEIR HARDIE (1856–1915)

One of the more far-reaching decisions that Liberal leader and future Prime Minister Henry Campbell-Bannerman [16] made, or tacitly approved of, in the run-up to the 1906 UK general election was that in certain parts of the country the Liberals would come to an agreement with the emerging Labour Party that one or other would stand aside in order to ensure that the Conservative candidate would be defeated. This decision gained the Liberals a few extra seats, although they would have won a landslide anyway, but crucially it allowed the Labour Party to win twenty-nine seats, a huge increase from their previous two seats, and gave Labour a significant presence in Westminster for the first time, an outcome that the Liberals would later come to regret at their political leisure.

The Labour Party that fought the 1906 General

Election was not actually called the Labour Party; since its founding in 1900 it had been known as the, rather less catchy, Labour Representation Committee. In an early example of New Labour rebranding, or to be exact New Labour Representation Committee rebranding, the party changed its name to the more straightforward Labour Party a week after the 1906 election and elected as its first leader James Keir Hardie, its most experienced MP and in the 1900s Britain's favourite leftie.

For Keir Hardie it had been a long road from where he was born in North Lanarkshire to the Houses of Parliament. Hardie came from a poor family, he had begun his working life when he was seven and by the age of ten was working down the mines. He had had no formal schooling but taught himself to read and write in the little time that he was not down the pit. His literacy made him an obvious spokesperson for his fellow miners and being a spokesperson who was willing to stand up for his colleagues saw Hardie lose his own job as a miner.

Undeterred, he turned to work as a union organiser and journalist before going into politics and, in 1892, became Labour's first ever Member of Parliament (MP) when he was elected by the good people of West Ham South in the working class East End of London. When Hardie turned up at the House of Commons he advocated previously unthinkable ideas such as, the rich having to pay more income tax, free education for all, state pensions for the elderly, the abolition of the House of Lords, devolution for Scotland and Ireland and women getting the vote; but what was really shocking was the fact that he eschewed the traditional MP's frock coat and top hat and, instead, wore a tweed suit and cloth cap – although to be fair it was a very smart tweed suit and cloth cap.

Keir Hardie helped found the Independent Labour Party in 1893 but by the end of the century had realised that, for the working class to have a truly effective organisation fighting on its behalf, the various socialist groups would have to combine with the growing trade union movement and, thus, in February 1900 the Labour Representation Committee was founded in London. To keep everything democratic and all the factions happy no leader was elected, although a young Scot by the name of Ramsay MacDonald [18] was appointed as its Secretary, but Keir Hardie remained the party's best-known figure and driving force and would, eventually, become leader of what – thanks to those generous Liberals in 1906 – would become the next big thing in British politics.

Albeit a reluctant leader, Hardie was a dedicated socialist, Christian and pacifist, forever fighting for equality and democracy until the end of his life. He was, of course, the sainted founder of the modern Labour Party, but was never to hold any governmental position – which was probably not too surprising as all that socialism, redistribution of wealth, abolishing the Lords and making the rich pay more was never going to catch on.

Powerful Scots

One must not forget that Scotland was, more or less, an independent nation for the best part of 600 years until the Union of the Crowns in 1603, even if for the most part this was down to the general indifference of, first, the Vikings and then the English who both had more lucrative countries to try and conquer. Scotland therefore became a kingdom on the fringes of Europe, with a history of perennial domestic squabbling, declining public interest and general international indifference, not unlike today's Scottish Premier League in fact.

Yet one must also remember that Robert the Bruce was the original Spiderman, and was it not Spidey who said that, 'with great power comes great responsibility' and for 400 years since 1603, Scots have found themselves holding positions of power and responsibility all around the world with, of course, the one notable exception being in Scotland itself as, much like the Romans, Vikings and English before them, Scots had come to the conclusion that it wasn't worth the bother.

REMEMBER, REMEMBER THE FIFTH OF NOVEMBER; REMEMBERING THE 30TH REMAINS OPTIONAL

11 JAMES VI (1566–1625)

For a nation that once ruled a quarter of the world the English have always had a propensity to allow themselves to be ruled by foreigners. First of all there was 300 years of Roman rule at the start of the first millennium, soon to be followed by the invasion from Germany of the Angles and the Saxons. The unified country of England would take its name from the Angles, but it would be the Saxons who would rule until the 10th century.

In the 11th century there was a period of twenty-five years when King Canute and his Danes conducted their early experiments in harnessing tidal wave power before, in 1066, the Normans from Normandy in France became England's conquerors. After the Normans, the next royal family to rule England was the Plantagenets of Henry

II in 1154 – who also came from France – and, finally, in 1485 when Richard III mislaid his horse at the Battle of Bosworth Field his successor was a Welshman, Henry VII, who began the canny House of Tudor. Therefore, it should not perhaps have been too much of a surprise that when Elizabeth I died in 1603 with no heir, the English looked northward to the House of Stuart – everybody else had had a go, so why not the Scots?

James VI left Scotland for London in 1603 where he was crowned king of both England and Ireland and became the first monarch to style himself King of Great Britain. He was the first Scottish king in more than 300 years to have been crowned on the Stone of Destiny, which of course now resided in London. James made no effort to return the stone to Scotland; it was an awful lot of bother to go to for some old piece of rock. In the remaining twenty-two years of his life James only went back to Scotland on one occasion, but considering that Scotland was where his father had been murdered, his mother had been forced to abdicate her crown and James himself as a teenager had been kidnapped and held hostage for a year, you can understand his reluctance.

James VI had been brought up as a strict Presbyterian but on becoming king of England in a high profile and highly lucrative transfer he moved to the Church of England and supported the imposition of Anglicanism throughout his kingdoms, much to the consternation of Presbyterians, Protestant Puritans and Catholics alike who had hoped for greater religious tolerance. On 5 November 1605 a Catholic plot to blow up the Houses of Parliament when James and the royal family would be present was foiled – an event that is somehow in these politically correct times still celebrated by communities around the country coming together to enjoy the spectacle of a Catholic being burned on top of a bonfire.

The events of November 1605 may well have had some bearing on James's difficult relations with the English Parliament and when he dismissed the house in 1614 because they did not agree with him he did not recall it until 1621. More to the point however, James believed in the Divine Right of Kings, which meant that the king was answerable to nobody other than God, and while his reign from 1603–25 was relatively peaceful and prosperous this policy would prove disastrous for the next Stuart generation.

The reign of James in England is known as the Jacobean era with Jacob being the Latin form of James, and is known today for the poetry of John Donne and the later plays of William Shakespeare including *Macbeth, King Lear* and *The Tempest.* This was the era when, from 1609, Protestant settlements, or Plantations, were introduced in Northern Ireland. Catholic estates in Ulster were confiscated by the state and further lands in Antrim and Down were acquired by Scottish landowners James Hamilton and Hugh Montgomery. Half of the original Protestant settlers came from Scotland and by the end of the 17th century further immigration saw the Scots become the majority Protestants and Protestants became the majority religious denomination. The Jacobean era also saw the foundation of the first British colonies in North America: in Virginia in 1607, in Newfoundland in 1610 and the Plymouth colony of the Pilgrim Fathers in Massachusetts in 1620 where a group of English Puritans fed up with turkey shortages and government restrictions on pointy hats set sail on the *Mayflower* in the ultimately misguided hope that there could be a world without stroppy Scots telling them what to do.

The first Colony was in Virginia and was named Jamestown after the King and Virginia became a thriving community through the profits of the tobacco industry. James was so concerned about the tobacco-smoking craze which was sweeping the nation that in 1604 he wrote *A Counterblaste to Tobacco*, although nobody paid any attention to it until 400 years later.

For all these momentous events, the name of King James is most associated with the version of the bible that he commissioned to be written in English by the greatest scholars in the Church of England, first published in 1611.

There was initially much resistance from non-Anglican Protestants to James's new, English version, but by the end of the 17th century Anglicans and non-Anglicans alike throughout Britain accepted it and over the following centuries would take the King James Bible to the rest of the world. This version remained the bible of all bibles for English-speaking Protestants until the 20th century when the temptation of free hotel room alternatives became just too much to resist. The bible first published in 1611 had become known as the Authorized Version or more commonly as the King James Bible, an astonishing historical accolade when you consider that James had of course not written a word of it.

WHERE'S YOUR HEAD AT?

12 CHARLES I (1600–49)

Who was the last king of Scotland? Idi Amin gave himself the title, King of Scotland, but no, it was not he, although the notorious Ugandan dictator did do some of his military training at Stirling and loved Scotland so much that he wore the kilt and gave Scottish names to four of his sons. The last monarch before the Act of Union in 1707 was Queen Anne, the last of the Protestant Stuarts and best known for her fine furniture but who never set foot in her northern kingdom. The last king of a separate and independent Scotland was of course James VI (1567–1603) [11] until on the death of Elizabeth he was proclaimed king of England and Ireland as well and ruled all three kingdoms until his death in 1625. However the last king to be born in Scotland was James's son Charles who on the death of his father succeeded the thrones of Scotland, England and Ireland.

Charles was born in 1600 in Dunfermline, Fife. Today, the town is best known for being the birthplace of one of the world's wealthiest men, Andrew Carnegie [26], and that of singer and actress Barbara Dickson, who sadly never knew either of her illustrious fellow Fifers so well. Dunfermline was an ancient capital of Scotland and the now ruined Dunfermline Palace where Charles was born was a royal residence from the 11th to the 17th century. Charles was the second son of James VI but when his elder brother Henry died in 1612 aged eighteen he became heir and in 1625 he became king.

Charles was very much his father's son and he, too, believed in the Divine Right of Kings, but where James through his years of having to survive the various factions of the Scottish nobility had learned to be pragmatic where necessary, Charles was impetuous, stubborn and would brook no counsel that he did not agree with. Furthermore while James's reign had been relatively successful, his disputes with the English Parliament and the Protestant Puritans lingered on and would become a lot, lot worse with Charles.

One of Charles's very first acts as king was to marry a Catholic princess, Henrietta Maria from France, which immediately got the Puritans' backs up. And by 1629 Charles's relations with Parliament had deteriorated to such an extent that he followed his father's footsteps by dissolving Parliament and ruling without it for eleven years. Not satisfied with alienating the Puritans and Parliament Charles set his sight on Scotland and made a concerted attempt to impose the Anglican religion and the Anglican Book of Common Prayer upon the Presbyterian Scots. This turned out to be provocation too far: the National Covenant of Scotland was signed in 1638 to protect the rights of the Church of Scotland and a Covenant army occupied the north of England to express Scottish opposition. Charles was forced to back down and, desperately short of money, he recalled Parliament in April 1640 with disastrous results – the so called Short Parliament only lasted three weeks. Six months on and still broke, Charles again recalled Parliament and disputes in this subsequent Long Parliament ended with the King walking into the House of Commons in January 1642 to arrest his opponents only to find that they had all fled: by the end of the year war had broken out between the King and the English Parliament.

For such a momentous period in British history, the civil wars of the 1640s are often given second billing to Henry and his wives or the reign of Good Queen Bess. Conflict came to Ireland and Scotland as well as England, but it was England that saw the most division with towns, neighbours and families having to decide which side to support. London and the east of England chose Parliament, with the Scots belatedly joining them. The west of England was predominantly on the side of the king who based himself at Oxford. Both sides had built up significant forces and the war would drag on indecisively for three years until the English Parliament forces combined with the Scottish army won an important battle at Marston Moor in 1644 and then, the following year, Oliver Cromwell's New Model Army won a decisive victory over the Royalists at the battle of Naseby. In 1646 Charles elected to surrender to his fellow Scots, who after much moral prevarication and Presbyterian angst about what to do with him handed their king over to the English Parliament, making sure that they were appropriately compensated financially for doing so.

Charles was then held under various forms of house arrest while the Parliament forces grappled with the dilemma of what to do with a king without a kingdom. If Charles had accepted his defeat graciously he might have been allowed to continue as a figurehead monarch, but Charles did not know the meaning of compromise and began to negotiate secretly with the same Scots who had handed him over in the first place. Charles proposed that he would support England becoming Presbyterian if the Scots would fight in his name. Charles of course had no intention of honouring this deal but, inexplicably, the Scots went along with it and invaded England in 1648 where they were easily defeated at the Battle of Preston.

This second civil war was the final straw for the hardliners on the English Parliament side and in January 1649 Charles was put on trial for high treason and found guilty. No matter how disastrous a king Charles had been it was generally thought that executing him was taking things too far but nevertheless, on 30 January 1649, Charles I, King of England, Scotland and Ireland, was led to the scaffold in front of the Palace of Whitehall and beheaded, although unlike his grandmother Mary, Queen of Scots, it only required one blow to remove his head.

After eleven years of the Commonwealth under the rule of Oliver Cromwell, the monarchy was restored in 1660 under Charles's son, Charles II. The two civil wars had resulted in the deaths of between 100–200,000 people and no matter what political or religious disputes remained outstanding nobody wanted to return to the battlefield to resolve them. These British Civil Wars have been regarded variously as wars of power, religion or of class and the ramifications would be felt in British history and society for centuries. Parliament had won and would become the dominant power, but the regret engendered by Charles I's very public death would ensure the role of the monarchy as Britain's first family. Future royals could be as mad and bad as they wanted, but at the first sign of turning dangerous they would immediately be exiled to France.

The Caroline era attracts far less acclaim in history than the Jacobean era, but it did see significant English expansion into North America; there were only 2,000 colonists in 1625 but by 1640 there were 40,000. In 1663 Charles II granted a charter for an English colony on the north-east American seaboard and it would be named Carolina in honour of his father. In 1712 the colony divided into North and South, and the Carolinas became two of the thirteen colonies originally under British rule that broke away to form the United States. During the 18th century, large numbers of Scottish Highlanders settled in North Carolina. When the American War of Independence broke out in 1776, while most Ulster Scots who had settled in the Carolinas fought on the Rebel side, ironically most of the Highlanders remained loyal to the Hanoverian British Crown, although true to form when one considers who the Carolinas were named after, it was the King's side who lost.

VIVA SOUTH LANARKSHIRE

13 THOMAS COCHRANE (1775–1860)

The rise of the British Empire in the 18th and 19th centuries was due in no small part to the supremacy of the Royal Navy. From time to time the Spanish, the Dutch, the French, or some combination of the three matched the British, but with the notable brief exception of the American War of Independence, British naval supremacy would always eventually be confirmed. The song *Rule, Britannia* celebrated this domination of the seas and ruling of the waves and became established as an alternative British national anthem beloved of the Last Night of the Proms, replete with cheering crowds waving Union Jacks. And the author of the lyrics of this most patriotic of British songs was as you have probably guessed a Scot.

James Thomson came up with the original words in the form of a poem in 1740. Thomson was born in the village of Ednam, near Kelso, which has the rare claim to fame of also being the birthplace of Henry Francis Lyte who wrote two of the world's most famous Christian hymns, *Praise, my soul, the King of Heaven* and *Abide With Me* and the latter has been sung at every English FA Cup Final since 1927. Lyte wrote *Abide With Me* in the knowledge that he had not long to live and would soon be meeting his Maker, adding extra poignancy to this most moving of hymns, although at least Lyte had the consolation that he would never have to witness Chelsea winning the final again.

The greatest challenge to British naval superiority was in the various conflicts from 1793–1815, which are referred to as both the Revolutionary Wars and the Napoleonic Wars, and an honourable mention should be made of Fife-born Frederick Maitland who, then a Royal Navy captain, happened to be in command of *HMS Bellerophon* in June 1815 off the coast of France when Napoleon attempted to flee France after his final defeat at the Battle of Waterloo. Maitland guessed correctly that Napoleon would attempt to sail from the port of Rochefort and blocked every attempt by the former Emperor to escape to America. Eventually on 15 July, 1815, Napoleon surrendered to Maitland and on setting foot on *Bellerophon* brought the Napoleonic Wars to an end.

With his honoured 'guest' safely aboard, Maitland set sail for England and anchored himself off Plymouth while the British authorities decided what to do with their bête noire. Maitland and Napoleon would dine together every evening and while relations between the two would remain cordial, the Scot Maitland politely rebuffed Napoleon's suggestions that wouldn't it be a good idea to resurrect the Auld Alliance.

Of course Britain's most famous naval hero is England's Admiral Horatio Nelson, but running him a close second, and more importantly keeping all limbs and eyes intact, was a Scotsman by the name of Thomas Cochrane from Hamilton. Cochrane came from a prominent Scottish family and joined the Royal Navy as a teenager. He made his name in the latter part of the Napoleonic Wars as a daring officer – the French called him the Sea Wolf – and, also, as a reformist MP at Westminster. Cochrane was as fearless in politics as he was at sea and gained many enemies in the establishment and in 1814 he was implicated in a fraud scandal, sent to prison for two years and kicked out of both the navy and Parliament. The disgrace would have finished most people, but Cochrane had the sea in his blood and if the Royal Navy didn't want him then he would find another navy that did.

In 1817 Cochrane was appointed commander of the newly formed Chilean navy in the country's fight for independence from Spain and, in conjunction with the revolutionary armies of Chile and neighbouring Peru, who were also fighting for their independence, attacked along the Spanish-controlled coast of Chile and Peru, gaining the new soubriquet of El Diablo to add to the Sea Wolf. By the time Cochrane relinquished command in 1822, both Chile in 1818 and Peru in 1821 declared their independence. But Cochrane was not finished with South America and in 1823 he took command of the Brazilian navy in their fight against their colonial masters, Portugal, who still controlled the northern coastal areas of Brazil. In 1823–24 Cochrane forced the Portuguese out of their remaining provinces of Bahia, Maranhao and Para, so securing the independence of Brazil and in Maranhao Cochrane surpassed himself by taking the province with a single warship and in recognition was appointed governor of the province. However Cochrane left Brazil in 1825 and continued his career of revolutionary-naval-commander-for-hire when in 1826 he joined the Greeks in their ultimately successful fight for independence from the Ottoman Empire. The Greek campaign was the only one in which Cochrane's involvement did not have a decisive impact but at least it improved Scottish-Greek relations, after Fife-born Thomas Bruce, 7th Earl of Elgin, removed a number of ancient marble sculptures from the Parthenon in Athens in 1806, in what turned out to be one of the more controversial early episodes of *Time Team*.

In view of Cochrane's complete success in achieving independence for all the countries he had so far been employed by, it was not surprising that the British authorities were keen for him to return to the fold before he offered his services to any country slightly closer to home and in 1832 he was reinstated in the Royal Navy with a full pardon and all his former titles restored. Sadly, the name of Thomas Cochrane is too often neglected today and the same is true of the large number of Scots who followed Cochrane to South America in the 19th century – with many settling as sheep farmers in Patagonia.

There are said to be more than 50,000 Chileans and Argentines today who can claim Scottish ancestry – the largest number of Scots descendants in any non-English speaking country. Of these the most influential was possibly Alexander Watson Hutton, a schoolteacher from Glasgow who settled in Buenos Aires. Hutton was a keen sports fan and in 1893 set up the Argentine Association Football League which would become the Argentine Football Association and so began the first football league outside Britain. Argentina would go on to become one of the major powers in world football, winning the World Cup twice and spawning such legendary footballers as Alfredo di Stéfano, Diego Maradona and Lionel Messi, and it is presumably this historic Scottish connection that saw so many Scots support Argentina versus England in the 1986 World Cup Quarter Final.

CHARGE OF THE BOYS' BRIGADE

14 GEORGE HAMILTON-GORDON, EARL OF ABERDEEN (1784–1860)

There was a significant difference in Anglo-Scottish relationships after the Union of the Parliaments in 1707 compared to after the Union of the Crowns in 1603. In 1603 James VI [11] and his Scottish entourage processed south to London to take up the reins of power, but in 1707 despite the Union purporting to be a full and fair merger between two countries to create the brand new nation of Great Britain, the reality was that all the power was retained in England with Scotland no more than a northern appendage and the Scots an irritation who had to be put up with. For most of the 18th century this remained the political reality and while Scots were given positions of power in the Colonies they were excluded at Westminster where the two political groupings during that century were the Whigs and the Tories.

Scotland did inadvertently provide the Whigs with their name. Both political parties' names were originally derogatory terms used during the British Civil Wars. Tory is derived from an Irish word 'toraidhe' meaning 'outlaw' and refers to when the Royalists were on the run from Cromwell's army in

Ireland. Whig came about following an event called the Whiggamore Raid in 1648 – w*higgamore* is Scots for 'cattle-drover' – when a group of radical Covenanters from Dumfries and Galloway rode to Edinburgh in protest against the possibility of the Scottish Government once more coming to the aid of Charles I [12] when they should be spending their time supporting local dairy farmers instead.

The term Whigs was later applied to those who supported parliament and Protestantism in preference to the monarchy and in 1714 when Parliament selected the German Hanoverians as the Stuarts' replacements on the throne, the Whigs became the dominant force in British politics for the next fifty years. However, when the young King George III succeeded to the throne in 1760 he decided it was time for a change and appointed friend of the family and his former tutor, Edinburgh-born Tory John Stuart, Earl of Bute, to become Britain's first Scottish prime minister in 1762.

The politically inexperienced Bute had been reluctant to take the position, and within a year had resigned and the Whigs returned to power. So traumatic had been Bute's short-lived premiership that, although it would be only a few years before the Tories regained control, it would be another ninety before a Jock was allowed anywhere near a position of power.

The next Scottish prime minister was another Edinburgh-born Conservative noble, George Hamilton-Gordon, who also happened to be the Earl of Aberdeen. He had begun his political career as a diplomat and then served as foreign secretary 1841–46. As foreign secretary he helped bring to an end decades of dispute between Britain and the US concerning boundaries in North America; after much heated negotiation the decision was made with the Oregon Treaty of 1846 to draw a straight line across the forty-ninth Parallel and that would fix the border between British North America and the US – that would later become the US/Canadian border – although Vancouver Island on the far west remained in British hands as the ruler they used to draw the line with wasn't quite long enough.

In 1852 Lord Aberdeen agreed to lead a coalition of liberal Conservatives and Whigs and was appointed prime minister. His reputation had been one of caution but in March 1854 he was persuaded to commit the cardinal sin of international politics – he declared war on Russia. Ever since the end of the Napoleonic Wars and British rule in India, Britain and Russia had been rivals for influence in Asia and the Middle East, but had so far managed to avoid direct conflict; however, when war broke out between Russia and the Turkish

Ottoman Empire in 1853 the British, with their new best friends the French, sent their navies to the Black Sea in support of the Turks. Even when the Russians agreed to withdraw from the lands they had occupied, both Britain and France went ahead and declared war. After all, they were already in the vicinity.

The Crimean War, so named because most of the fighting took place on the peninsula of Crimea, in what is now the Ukraine, lasted until 1856, and was a conflict which became known for military incompetence on all sides and which was exemplified most famously at the Battle of Balaclava in 1854. This battle would in normal circumstances be remembered for the brave action by the Highland Brigade under the command of Colin Campbell from Glasgow, who three years later would lead the British response to the Indian Mutiny, or to give its correct name the First Indian War of Independence. At Balaclava, and standing only two-deep, the Highland Brigade successfully repulsed an attack by more than 2,000 Russian cavalrymen and the stand they took would later become immortalised in the phrase, the Thin Red Line – the Highlanders wearing their red uniform jackets and stretched thin along the battlefield.

But the Battle of Balaclava was no normal battle as it also featured the infamous Charge of the Light Brigade when the British light cavalry were mistakenly sent on a suicidal attack of the wrong Russian position. So reckless was the action at Balaclava that the Russians thought the British must be drunk, the French thought the British must be mad, and knitted headwear manufacturers thought they should probably make the holes for the eyes a little bigger.

The Crimean War is considered the first modern war in that, for the first time, the conflict was photographed and through the advent of the telegraph, newspaper journalists could quickly report back on the latest military debacle as well as highlighting the terrible conditions for the soldiers, with far more dying of disease than in battle; only Florence Nightingale and lamp manufacturers managed to come out of the conflict with their reputations enhanced.

The reports in British newspapers increased the unpopularity of the war at home and Lord Aberdeen was forced to resign in February 1855 and thereafter retired from politics. The Crimean War came to an end the following year with more than 300,000 dead in total and, although Britain was nominally on the winning side, within twenty years the Russians had returned to the Black Sea and the Ottoman Empire continued its long decline into oblivion. A comprehensive post-Crimean War inquiry would establish that never under any circumstances should you ever go to war against

Russia, although the inquiry's second major proposal, that it is not a good idea to go to war in the Middle East for no good reason, would sadly be eventually forgotten.

DID YOU REALLY MEAN TO SEND THAT?

15 ARTHUR BALFOUR (1848–1930)

The third of Britain's Scottish-born prime ministers was, like the first two, a Conservative, although unlike Lord Bute and Lord Aberdeen, Arthur Balfour from East Lothian was not a peer, but he was certainly very well connected. Balfour's father was an MP and his uncle on his mother's side was Robert Gascoyne-Cecil, Marquess of Salisbury, and three times Conservative prime minister of Britain between 1885–1902. It was therefore no surprise which career young Arthur chose to follow and he was elected as an MP in 1874, became Gascoyne-Cecil's parliamentary private secretary and first served in his uncle's Cabinet in 1886. With the Marquess of Salisbury sitting in the House of Lords, Balfour became Conservative leader in the Commons in 1891 thus continuing his apprenticeship as heir apparent to lead the party; the maxim, Bob's your uncle, is said to be coined in his honour (although one suspects that even Balfour would never have addressed the noble Marquess as Uncle Bob).

When Salisbury retired in 1902, all his patronage

proved to have been worthwhile as Balfour was appointed party leader and, therefore, prime minister. His elevation had been unchallenged but his premiership was marked by internal division between Conservative party members who supported free trade and those who supported protectionism – the split led to the resignation of Balfour's government at the end of 1905 to be replaced by a Liberal government, led by fellow Scot Henry Campbell-Bannerman [16].

The main achievement of Balfour's government was the entente cordiale of 1904 which marked a new beginning in relations between Britain and France after centuries of enmity and rivalry, although lingering suspicion about each other's cuisine resulted in the celebratory official banquet consisting of a neutral Chinese take-away.

Balfour remained Conservative leader until 1911, losing three successive general elections to the Liberals in the process. And despite being a rather insular and pessimistic soul who once said memorably that, 'Nothing matters very much and most things don't matter at all', he spent an inordinate amount of time from 1906–11 leading – in conjunction with the House of Lords who at the time had the power to repeatedly block legislation from the Commons – in a concerted rearguard action to prevent as many of the Liberal's progressive reforms as possible. This obstruction would eventually lead to a constitutional crisis which was resolved only with the Parliament Act of 1911 which finally established the supremacy of the Commons over the Lords, as by this stage even the new king – George V, crowned in 1910 – had to concede that the plebs could not be held back forever.

Balfour returned to government in 1916 as foreign secretary in a wartime coalition and he would serve in this post for the remainder of World War I and until 1919. And he would return, again, in the 1920s as a Conservative elder statesman, despite, or perhaps because of, his patrician, aloof manner and a reputation for procrastination: it was once said of him, 'If you want nothing done – then Balfour is your man'.

For all Arthur Balfour's involvement at the forefront of British politics for more than forty years, he is remembered today for the Balfour Declaration – a one page letter that he wrote while foreign secretary. On 2 November 1917, Balfour, representing the British government, wrote to Lord Rothschild who represented the Zionist Federation of Britain and Ireland as follows:

> His Majesty's Government view with favour the establishment in Palestine of a national home for the Jewish people, it being clearly understood that nothing

> should be done which might prejudice the civil and existing rights of the non-Jewish communities in Palestine.

In 1917 less than ten percent of the population of Palestine was Jewish, but this had not stopped Jewish organisations and the First Zionist Congress calling for a homeland in Palestine from 1897 onwards. In 1906 Balfour, who was at that point an MP in Manchester, met one of his constituents, Chaim Weizmann – a chemistry lecturer at Manchester University and leading British Zionist – and was persuaded of the merit of a Jewish homeland in Palestine. Eleven years later, Balfour was in a position to do something about it when, on behalf of the Government, he sent the letter which would have such dramatic consequences for 20th-century political history. Even though there has been some debate about how many of the British government actually supported what Balfour was proposing, in late 1917 Britain was still very much in the middle of World War I, a conflict they still had every prospect of losing, and were busy making promises to all and sundry in the hope of gaining support. As well as promising a Jewish homeland in Palestine, the British were also telling the Arabs that they would gain their independence if they rose up against the Turkish

Ottoman Empire, who ruled the Middle East including Palestine and were on the opposing side to Britain. While all this was going on the British had secretly agreed with the French that if they somehow did manage to win the War, the two colonial powers would take over the Middle East and divide it between them.

In the December of 1917, the British captured Jerusalem from the Turks and when World War I was finally over in November 1918 the Middle East was divided between the British and the French. Both powers administered their spheres of influence under the mandate of the League of Nations with the vague assumption that the Arab former Ottoman-run territories would eventually gain their independence and that at the same time this would somehow also include, as promised, a national home for the Jewish people.

The British drew the short straw that broke the camel's back and ended up with Palestine. Faced with having to deliver on all their various and conflicting promises the British came up with a cunning plan of dividing Palestine in two, giving seventy-five percent of the land to a new Arab kingdom of Jordan and retaining the twenty-five percent of Palestine in the west in the vain hope that if they gave it long enough then they could eventually figure out what to do with it. The British remained in Palestine until 1948, by which time more than 400,000 Jews had immigrated to the Holy Land, but they still comprised less than one third of the total population of Palestine. Arthur Balfour, by this time Lord Balfour, visited Palestine in 1925, and while some of the Jews gave him a warm welcome, as far as both the Muslims and the British governors were concerned it was just another reminder that life would have been so much easier if, instead of writing his Declaration, Balfour had followed the maxim which had stood him in such good stead throughout his political career and done absolutely nothing.

Unable to solve the problem entirely of their own making, the British gave up and left Palestine in 1948, resulting in civil war and the creation of the State of Israel, with Balfour's old friend from Manchester, Chaim Weizmann, becoming the new nation's first president. More by accident than design a Jewish homeland in Palestine, as promised by Arthur Balfour, had come to pass and in the process one million Palestinians lost theirs, and the part of the Declaration that stated, '... nothing should be done which might prejudice the civil and existing rights of the non-Jewish communities', was completely forgotten. The Balfour Declaration has become one of the most controversial letters in world history, celebrated by some as a crucial moment in the creation of modern Israel and criticised by others as one of the worst mistakes in Western foreign policy, although if it is any consolation to Arthur Balfour, when it came to the Middle East, the following century would produce no shortage of contenders for that particular unwanted accolade.

MENZIES CAMPBELL KNEW MY GREAT-GREAT-GREAT-GREAT GRANDFATHER

16 HENRY CAMPBELL-BANNERMAN (1836–1908)

It may come as a surprise to all those who had never heard of the Liberal Democrats before Nick Clegg won the first round of *Britain's Got Political Talent*, but the Liberals have a long and once illustrious history. The man who created the modern Liberal Party, who was four times prime minister, the Grand Old Man of British politics (or as his great rival Benjamin Disraeli nicknamed him, 'God's only mistake') and son of a Leith merchant, William Ewart Gladstone had died in 1898 at the age of eighty-eight. All Gladstone's likely heirs had either retired or resigned and the Party was internally divided, in opposition, and with little prospect of a swift return to power, so little was expected of the new Liberal leader who took over in 1899.

Henry Campbell-Bannerman was born in Glasgow,

his father was Lord Provost and his family had made their money in the drapery business. He became MP for Stirling in 1868 and had held several Cabinet positions, but by 1899, the year he became leader, he was something of a grand old man himself, although with his family background a rather dapper one.

Campbell-Bannerman had been elected as the compromise safe-pair-of-hands candidate who would mind the shop until a more promising leader emerged, but he ended up in charge for a decade. His great strengths as a politician were his consistency and his ability to get on with all sides of his party – after all those years working in a draper's shop he had a knack of telling people what they wanted to hear. In 1905 it was the Conservative Government led by fellow Scot Arthur Balfour [15] which ended up breaking down amid internal divisions and when Balfour resigned, King Edward VII called for Campbell-Bannerman, as leader of the opposition, to form a new government. Campbell-Bannerman became prime minister in December and he immediately called for a general election to give himself and his party a fresh mandate – a lesson that another appointed new Scottish prime minister a century later would have been wise to follow.

The Liberals won the 1906 General Election by a landslide, gaining a 125-seat majority, and began enacting a system of social reform which laid the ground for the beginning of the welfare state system with the introduction of old age pensions, free meals and medical inspections for school children, job centres for the unemployed and greater rights for workers and trade unions. The Liberal government of 1906 would become one of the most important in modern British history and it was Campbell-Bannerman who would become the first person to be officially given the title of Prime Minister, as although the title had been in common use for many years, before Campbell-Bannerman the resident of No 10 had to make do with the official title of First lord of the Treasury.

Sadly, Henry Campbell-Bannerman would not live to see through this historic parliamentary term; he died in 1908 and was succeeded by two of the most prominent members of his Cabinet, Herbert Henry Asquith who was prime minister from 1908–16, followed by David Lloyd George who held the office from 1916–22. Asquith and Lloyd George are far more readily remembered today than is Campbell-Bannerman but, nevertheless, neither was ever able to emulate their predecessor's landslide victory of 1906, which remains to date the zenith of British Liberalism when Liberals were the kings of British politics and the instigators of a modern and progressive state rather than merely the king-makers who get to make all the cuts.

OH WHAT A LOVELY POPPY FIELD

17 DOUGLAS HAIG (1861–1928)

It was not Douglas Haig who was responsible for the massive arms race taken up by the major European powers in the decades leading up to 1914 and it was not Douglas Haig who was responsible for the political machinations, alliances and intrigues which made war to various prime ministers, presidents, kaisers, tsars, kings and munitions manufacturers not just inevitable but also desirable. And it was not Douglas Haig who took such exception to Franz Ferdinand's disappointing second album that he decided to assassinate the Archduke in Sarajevo in 1914 and kicked the whole thing off.

World War I, the War to End All Wars, is possibly unique in being almost bereft of any semblance of a hero, but every war has to have a villain or two and as far as the British and the British Commonwealth are concerned it was not King George V, in whose name the war was being fought , nor was it the two British prime ministers, Asquith from 1914 to 1916 and Lloyd George from 1916 to 1918, who were in charge of war effort and strategy. That unwanted accolade has been given instead to the British generals who oversaw the carnage and a loss of life in battle that was beyond anybody's comprehension, but in particular to Edinburgh-born Douglas Haig who became commander-in-chief of the British armed forces in December 1915 and at the Battle of the Somme saw 600,000 Allied casualties – alongside 600,000 German casualties – for the great achievement of pushing back the German army the grand total of seven miles.

It was the German High Command who are supposed to have coined the phrase that the British army were 'lions led by donkeys', and one suspects that they had the obstinate, single-minded Field Marshal Douglas Haig very much in mind when they said it. Haig came from the famous Haig whisky family, but sadly after studying at Oxford young Douglas decided that a life in the distillery was not for him and enrolled at Sandhurst and began his career as a soldier. Haig saw action in the Boer War and then assisted Liberal government minister Richard Haldane, also from Edinburgh, in the reforms that restructured and modernised the British army and saw the creation in 1908 of the British Expeditionary Force, a British army formed for the specific purpose of fighting in a war overseas – just for emergencies of course. When war broke out in 1914 Haig was given command of the British First Army Corps and by December 1915 with the war already bogged down in trench warfare and horrendous casualties, Haig had finally grown a moustache rigid enough to be appointed overall commander of the British armed forces and remained so until the end of the war in November 1918.

Douglas Haig is most associated with two particular battles: the Battle of the Somme, July–November 1916, and the Battle of Passchendale (or as it is often known the Third Battle of Ypres) from July–November 1917. The Somme is a river in the north of France and although the battle there was the first in which tanks had been used in action it is far better known for its excessive loss of life: on the very first day, 1 July, when Haig sent his men over the top, more than 19,000 were killed and a further 40,000 were wounded, which meant that twenty percent of the entire British and Commonwealth army had been killed or wounded in less than twenty-four hours – and all of them volunteers.

Such unprecedented slaughter caused even the most blinkered to pause and consider, not as you might expect whether the war should be continued at all, but whether Haig should remain in post to continue his policy of what became known as offensive attrition, i.e. we'll keep going no matter what the casualties are. However, the British government had abdicated much of its responsibility for the running of the war to the army and the other generals when surreptitiously approached about replacing Haig were more than happy to let him take all the blame, as to be honest they didn't have a plan B either.

Thus, a year after he launched the disastrous Somme offensive Haig tried again, but this time in Belgium with the Battle of Passchendale in which it took some four months of fighting to capture the small Belgian village. Even more than the Somme, Passchendale symbolises the futile slaughter and human misery inherent in the conflict on the Western Front with the deaths of more than 300,000 Allies and 250,000 Germans and a battlefield so deep in mud that tanks were useless. When Passchendale was taken in November the Allies had gained five miles, which when added to the seven miles gained at the Somme meant that for a total of just under one million casualties the Allies had gained twelve miles of mud – and would proceed to lose it again only a few months later in the German spring offensive of March 1918.

Douglas Haig's strategy had been that, with two such evenly matched and heavily armed forces, a war of attrition was the only way to go. The combined manpower of Britain, France and Russia, he reasoned, would always win over Germany, Austria and Turkey because as long as the Allies were killing the Central Powers at roughly the same rate as they were killing us, the Germans would run out of men first. And for the three years Haig was in charge that was the way it was. Defenders of Haig have pointed out that his strategy proved ultimately correct, as the Allies did eventually win, but it was a perilously close run thing: with the Russians otherwise engaged in various revolutions in 1917 and therefore out of the remainder of the War, the Germans were only fighting on one front and nearly broke through in the spring of 1918, but the Allies – bolstered by US troops – held firm and on 11 November 1918, the guns fell silent. Germany had been defeated and, four years after they had been promised, the lucky few survivors would be home for Christmas.

Douglas Haig remains a highly controversial figure to this day. After the war he devoted himself to working for ex-servicemen through founding the Earl Haig Fund and in his role as the first president of the British Legion in 1921 he helped organise both Remembrance Sunday and the Poppy Appeal. Haig is also given credit for his resolute defence during the German offensive of spring 1918 and his famous Special Order of the Day, 'With our backs to the wall and believing in the justice of our cause each one of us must fight on to the end' – from which comes the adage, backs to the wall. Haig, however, ensured that he was always a safe number of miles behind said wall.

But despite his considerable post-war charity work and his single-minded determination throughout the war to achieve victory, regard for Haig must always be tempered by his diary entry on 1 July 1916, the first day of the Battle of the Somme,

'. . . the total casualties are estimated at over 40,000 to date. This cannot be considered severe in view of the numbers engaged . . .'

IT'S MY PARTY AND I'LL LEAVE IF I WANT TO

18 JAMES RAMSAY MACDONALD (1866–1937)

Until such time as Tony Blair [20] wins a Nobel prize for his unconventional attempts to find peace in the Middle East, there have been two Scots' recipients of the Nobel Peace Prize. In 1949 renowned nutritionist John Boyd Orr from Ayrshire won the award for his tireless work from the 1930s onwards that showed that poverty and poor diet were linked to ill health and became an integral part of Britain's wartime rationing policy. After the war Boyd Orr became the first director general of the United Nations Food and Agriculture Organisation and spent the remainder of his life until 1971 campaigning against world hunger.

The other peace prize recipient was Glasgow-born

Labour politician Arthur Henderson, who served as both home secretary and foreign secretary, and won the prize in 1935 for his chairmanship of the World Disarmament Conference held in Geneva between 1932–34 which had as its magnificently ambitious ideal the reduction and limitation of offensive weapons so that there would never be war again. Sadly, as with all international conferences, most of the major powers didn't really believe in the ideals they claimed to espouse and it became clear that the Geneva Conference was bound to fail. But at least the international community made sure that Henderson won a consolation prize for all his efforts; a symbolic gesture of peace and hope acknowledged around the globe – at the same time as the world was rearming as fast as possible.

Arthur Henderson was a major politician in the British Labour Party for more than thirty years and would lead the party three times between 1908–32. Indeed, all the party's first five parliamentary leaders were born in Scotland – Henderson, the sainted Keir Hardie (10) and, in the early 1900s, George Barnes from Dundee and William Adamson from Dunfermline, but in terms of political importance James Ramsay MacDonald from Moray wins out. MacDonald would be the dominant figure in Labour's rise to become one of the two major parties in 20th-century Britain.

James Ramsay MacDonald was born in the fishing town of Lossiemouth and his parents, a crofter and a servant, were not married. He studied hard at school and worked as a teacher before moving to Bristol in 1885 where he first became involved in radical socialism and then to London in 1886 where he found a position as a clerk. In 1896 MacDonald married Margaret Gladstone (no relation to the four-times prime minister, but great-niece of Lord Kelvin) whose money provided an income which allowed MacDonald to concentrate on a life in politics and when Keir Hardie founded the Labour Representation Committee (the Labour Party's original name) in 1900, MacDonald was elected Secretary. It was as Secretary that MacDonald negotiated the electoral pact with Henry Campbell-Bannerman's [16] Liberals which resulted in the Conservative defeat at the 1906 General Election and gave Labour its first major electoral breakthrough. Conveniently, one of the constituencies the Liberals agreed not to contest – so that Labour's chance of a win was improved – was Leicester, where MacDonald just happened to be standing.

Ramsay MacDonald became Labour leader in 1911, but was forced to stand down in 1914 due to his opposition to World War I. For his stance, MacDonald would be vilified as cowardly, unpatriotic and worse –

his illegitimacy often cited pejoratively. His political rehabilitation and return as party leader would not be complete until 1922 by which time most of the country tended to agree that, in hindsight, the Great War had not been all that great. Britain in the 1920s was not the 'land fit for heroes' that people had been promised, and when the Liberals finally delivered universal suffrage to men over the age of twenty-one and women over thirty the working class thanked them by voting Labour in the general election of 1923 and Labour, with 191 seats, pushed the Liberals into third place. Now, as the second party in the House of Commons and with the Conservatives unable to form a working government, MacDonald was asked to form a minority administration and in January 1924 he became Britain's first Labour prime minister, Britain's first working-class prime minister and, with a nod to Lossiemouth, the first prime minister to know the precise price of fish.

This first Labour government would last only ten months – minority governments were not as popular as they are today and the Conservatives were returned to power within the year. However, despite being attacked mercilessly by the right-wing press, MacDonald had ensured that the country's boat had not been rocked, the horses had not been scared and the aristocracy had not been stood up against the wall and shot, and in the

1929 General Election Labour defeated the Conservatives for the first time winning 287 seats to the opposition's 260. MacDonald was once more prime minister, although again requiring cross-party support to govern. However, Ramsay MacDonald had achieved his lifelong ambition, the ultimate goal of Labour becoming the largest party in the country – even if, as appears to happen with most Labour heads of government, he had pragmatically dropped most of the radical socialism he had previously espoused in the process.

Unfortunately the 1929 Government coincided with the start of the Depression in Britain, and thus began the long tradition of Labour administrations running out of money – these days such an eventuality would be solved by borrowing trillions and letting someone else sort out the mess, but back in the early '30s they decided that they should attempt to balance the books. The Government was faced with the unpalatable choice of having to make huge cuts in public expenditure, including reducing unemployment benefit in the face of steeply rising unemployment. MacDonald supported the cuts but was a step too far for most of the Cabinet and he felt obliged to offer his resignation in August 1931.

If that had been the end of the story and the end of

MacDonald's political career it would probably be he, rather than Keir Hardie, who would have become the hero of British socialism. But, sadly, Ramsay MacDonald liked being prime minister and was persuaded by King George V to continue as prime minister and form an emergency National Government that included representatives from the Conservatives and Liberals and which would enforce the spending cuts. The majority of the Labour Party wanted nothing to do with this; MacDonald was expelled from the party he had led to power, and this time there was no chance of a comeback. MacDonald decided to call a general election in October 1931 to gain a mandate for his new government and was rewarded with a colossal 556 MPs elected under the National banner, although 473 of these were Conservatives, and those who were left in the Labour party were decimated and left with only 52. MacDonald remained prime minister until his resignation through ill health in 1935, but he commanded little power – sidelined by his Conservative 'allies' and reviled as a traitor by former Labour colleagues.

MacDonald's incredible achievements in building up the Labour Party that he had helped found in 1900 and had seen rise from no more than 2 MPs to become the party of government are but nothing compared to his crime of sleeping with the enemy. His life and career spell a lesson that no future Labour politician would ever forget, and that should give pause to all Liberal Democrats: you can be as right wing as you wish, but you never ever join the Tories.

NEVER WAS SO MUCH OWED BY SO MANY TO SOMEONE CALLED HUGH

19 HUGH DOWDING (1882–1970)

When twenty-one years after the War to End All Wars ended in 1918, World War II began, there was to be for good or for ill no Scottish equivalent of Field Marshal Douglas Haig [17]. If many of the lessons of the first war had been forgotten long before 1939, one that had been remembered was that politicians should not allow military leaders the power and latitude to follow their own strategy regardless of consequence. From World War I we remember the generals and the poets, but in World War II it would be the dictators, the presidents and the prime ministers.

There would however be one Scots-born Prime Minister during World War II and that was Peter Fraser from near Tain in the Scottish Highlands, who was Prime Minister of New Zealand from 1940–49. During 1940–45 more than 140,000 Kiwis served abroad in the 2nd New Zealand Expeditionary Force seeing action in

Greece and North Africa before ending the war in a whirlwind campaign across northern Italy.

The leading Scottish-born British politician during the second war was John Anderson who hailed from Edinburgh and served as both home secretary and Chancellor of the Exchequer. As home secretary he is remembered for being responsible, prior to the outbreak of war, for home defence and air raid precautions that included the development of what would become known as the Anderson shelter. The shelter was made of corrugated iron sheets and measured 6 ft (1.8 m) high, 6 ft 6 in (2 m) long and 4 ft 6 in (1.4 m) wide. From 1939 the pressure was on to provide air raid protection and more than three million shelters were distributed to households throughout the country. The Anderson shelter proved so robust that decades later they were still being used in allotments.

By June of 1940 German invasion seemed imminent, France, the Netherlands, Belgium, Norway and Denmark had all been defeated by Germany, and with the Russians and Americans still neutral, the only hurdle left to jump before Hitler achieved a final victory was the aerial domination necessary to invade Britain. Herman Goering's so far invincible Luftwaffe had 4,000 aircraft – double that of the Royal Air Force (RAF). The Battle of Britain began on 10 July 1940 and lasted until the end of October and leading RAF Fighter Command and Britain's aerial defence was Air Chief Marshal Hugh Dowding from Moffat in Dumfries and Galloway.

Dowding had moved to England with his family as a boy. He joined the army in 1899 and after gaining his pilot's licence in 1913 joined the Royal Flying Corps (RFC) that had been founded the year before with Glasgow-born David Henderson as its first commander. Dowding served in the RFC in World War I and then with (as the RFC became in 1918) the Royal Air Force and by 1936 had been promoted to become the first commander of the newly renamed Fighter Command.

He spent the remaining pre-war years ensuring that Britain was as prepared as possible for attack from the air. He established an integrated defence system of communications, observers and radio; he introduced the Spitfire and Hurricane airplanes and, crucially, he promoted the introduction of a revolutionary new detection technology, radar, which was being developed by Scottish engineer Robert Watson-Watt [100]. Dowding had been due to retire in 1939, but when war was declared he remained in command.

Over the English Channel and the English countryside the Luftwaffe threw everything they had at Fighter Command, attacking the airfields, radar stations and airplanes, but the RAF put up stiff resistance and the enemy suffered heavy losses. Dowding, despite having considerably fewer aircraft than his opponent, had the advantages of home territory and a defence system, including radar, which he himself had put in place. Dowding aimed always to keep planes and pilots in reserve but there was a limit and if Goering had kept up his attacks then by dint of its superior numbers the Luftwaffe may well have won the day. However, at the end of August the Germans switched their strategy to the Blitz and night time bombing which allowed the RAF time to regroup, build more planes and train more pilots and by October 1940 Hitler had reluctantly decided to postpone invasion until the spring of 1941 at the earliest and, by then, priorities had changed and the Nazis were going on a bear hunt.

In total, the RAF lost more than 600 planes in the Battle of Britain, but the Germans lost around 1,200. Germany had suffered a serious setback and Britain and her Commonwealth Allies would live to fight another day and it gave a sign to the watching Americans, who always like to be on the winning side, that maybe the Limeys had a chance after all.

In a speech to the House of Commons in August 1940 Prime Minister Winston Churchill – in office since May 1940 – famously said that, 'Never in the field of human conflict was so much owed by so many to so few . . .', and then did what most politicians normally do and sacked the very person who led the few to whom so much was owed. Dowding was not the most diplomatic of men and now that the immediate danger was over Churchill, and his other superiors, could finally afford to let him go. Understandably Dowding was not especially happy about this dismissal one month after his finest hour and remained extremely bitter about his treatment for many years.

Three years after Hugh Dowding learned to fly in 1913, another Scot, Arthur Tedder from near Loch Lomond joined the RFC and he, too, survived the aerial dogfights of the first war and continued his career in the RAF. In 1942 Tedder was promoted to Air Chief Marshal and led the RAF in the Mediterranean and North Africa where he supplied decisive aerial support in the allied victory at El Alamein. In 1944 Tedder was appointed Deputy Supreme Commander under Eisenhower for the preparations for Operation Overlord and the Normandy Landings.

D-Day had been planned originally for 5 June 1944 and 175,000 Allied forces, 50,000 vehicles, 11,000 aircraft and 5,000 ships were marshalled and ready to cross the Channel. The Germans knew they were coming – but not when or where their enemy would land. In early June the weather turned and with rain and storms in the English Channel the original date was abandoned.

Eisenhower was in two minds whether to go ahead on June 6 or postpone indefinitely. His senior staff were also divided about what do, but Eisenhower had to make his decision no later than the evening of the 5th and thus consulted his senior meteorologist James Stagg from Dalkeith in Midlothian. Stagg reported that the weather would remain unsettled and stormy for the next few days but there would be a brief, few hours' break the following morning. And on the basis of this only known example of a Scotsman giving an optimistic weather forecast Eisenhower decided that the D-Day landings, the largest invasion by sea the world has ever known, would begin at 0630 hrs on 6 June 1944.

FORTY-FIVE MINUTES IS A LONG TIME IN POLITICS

20 TONY BLAIR (1953–)

Ah Mr Blair, we were expecting you, but where to begin? Perhaps we should start with the only other British prime minister to serve for more than a decade in the past century and the Prime Minister to whom Tony Blair is most often compared. Of all the many political consequences of Margaret Thatcher's premiership from 1979–90 one of the most far-reaching was the increasing political division between England and Scotland. Thatcher was a dyed-in-the-wool Unionist who may once have truly believed that her policies, 'would bring harmony where once there had been discord', but soon this had been transformed into those who are, 'one of us' and, 'everybody else'. And while Middle England happily signed up to join the former camp, contrary, recalcitrant, Presbyterian Middle Scotland increasingly found themselves bridling at being told when and where to rejoice and joined the latter camp – having first made sure they had bought their house, their share portfolio and set up their private pension plan of course. Therefore, while in England the Labour Party was trounced in the 1983 and '87 General Elections, it remained relatively intact in Scotland and became the main benefactor of the growing anti-Thatcher consensus among those who clung to the traditional, socialist belief that the state would and should provide and its new-found allies who just wanted shot of That Woman.

The Labour opposition in Westminster now had a disproportionately high number of MPs from Scotland and Scottish MPs were thus beginning to take on increasingly prominent roles in the higher echelons of the party leadership. The most senior of these Scots was John Smith from Argyll who became leader in July 1992, the first Scot to hold that position since 1932. Had John Smith been Labour leader at the May 1992 General Election two months earlier it is possible that he would have become Prime Minister and had he been Labour leader at the 1997 General Election he would definitely have become Prime Minister. Sadly, a Smith premiership has become one of the great 'what ifs' of British politics since his death from a heart attack in May 1994 resulted in an unexpected vacancy for the job. Gordon Brown [30] was the early favourite to succeed – he was already in charge of Labour's economic policy – but with what appeared unseemly haste support switched swiftly to another, younger, more photogenic Scot, who as far as those all important English marginals were concerned had the added advantage of not sounding Scottish in the slightest.

Tony Blair was born in Edinburgh, spent part of his childhood in Glasgow and went to the exclusive Fettes independent school in Edinburgh. Blair also spent part of his childhood in Australia and the north of England before graduating in law from Oxford University. He became a Labour MP in Sedgefield, Durham in 1983 and was promoted to the Shadow Cabinet in '88. He and Gordon Brown were the new generation of Labour politicians and in '94, during a legendary meeting at Granita, a trendy London restaurant, the two men agreed at the end of the meal that whoever was closest to guessing the final bill would stand for leader, while the other would stand aside. Brown then proceeded to tot up on a napkin various complicated sums, but typically came up with a considerable overspend, while Blair, who ate in the restaurant regularly, came up with the exact figure as he believed it to be right.

Tony Blair was elected Labour leader in '94 and won the general election in June '97 with a landslide majority of 179 seats – Labour's greatest ever parliamentary majority and greatest number of seats won. At the age of forty-three Blair was the youngest British prime minister since 1812 and the first Scot to hold the office since the unlamented Ramsay MacDonald [18]. Under Blair's leadership Labour would win two further general elections, in 2001 and '05. Tony Blair remained in the top job until his retirement in June 2007 making him the longest-serving Labour prime minister in history, a fact that stuck in the craw of many longstanding party supporters who never knew what Mr Blair and Labour ever had in common in the first place.

Blair took Labour to the free-market economy centre ground safe in the knowledge that those on the left would have nowhere else to go. Socialism and public ownership were out, and New Labour, the New Millennium and a New Dawn were in – although in a sign of things to come Blair's commitment to a new dawn was immediately broken on election night.

As part of their 1994 agreement, and under the

constant threat of having that effin grin wiped of his face if he ever tried to wriggle out of it, Blair left economic issues to new Chancellor of the Exchequer Gordon Brown. The British domestic economy was booming through rising property prices, high consumer demand, and a buoyant financial services sector, the minimum wage and civil partnerships were introduced and in the late '90s the new, young, charismatic PM was at the forefront of a Cool Britannia that lasted until Oasis's third album was released and the Millennium Dome was opened.

Constitutionally, Blair's government was responsible for the biggest changes in the UK for eighty years: the Good Friday Agreement of 1998 paved the way for the intermittent return of self-government in Northern Ireland and '99 saw the creation of the Welsh Assembly and the restoration of the Scottish Parliament. It never seemed clear how personally enthusiastic Tony Blair was for Scottish devolution – or for the land of his birth in general – and there was always suspicion that with an overwhelming parliamentary majority behind him he was of the opinion that a Scottish parliament might prove more trouble than it was worth. However, with Labour far and away the most popular party in Scotland and with proportional representation imposed to ensure that the Nationalists could never win power what could possibly go wrong?

With Gordon Brown minding the shop at home and now that the Northern Irish, Scots and Welsh had been sorted out, and with the People's Princess out of the way, the time had come for a self-confident Tony Blair to take his rightful place on the international stage.

In March 1999 he was one of the principal voices persuading the US in particular and NATO in general to intervene in the war in Kosova with the aim of stopping the military conflict between the majority Albanian population and the Serbian government and people. From March to June that year NATO carried out sustained bombing on Serbian targets that would eventually force the Serbs to withdraw, allow United Nations (UN) peacekeepers to move in and the people of Kosova to return to their homes. Although this intervention was generally seen as a good thing, and within a year the not very nice Serbian leader Slobodan Milosevic would be forced out of office, in a portent of what was to come NATO had acted without the support of the UN Security Council. In April '99, in the midst of the bombing, Blair made a speech in Chicago in which he espoused his belief that the international community had a right to intervene militarily in conflicts where the civilian population of that country were at risk. This policy became known as the Blair Doctrine in order to differentiate it from the film *The Blair Witch Project* of the same year, although it was unclear which one would eventually prove scarier.

Tony Blair became famous for his ability as an effective communicator and his easy rapport with those that he worked with. Having built a close relationship with US President, Bill Clinton, he did the same with Clinton's successor George W. Bush. In 2001 when the Americans, with Bush at the helm, decided to retaliate in the aftermath of 9/11 and attempt to topple the Taleban in Afghanistan, Blair was the first to sign up to join the invasion, little suspecting that the Taleban, the Americans, the British, the opium fields and Osama bin Laden would all still be there nine years later.

If Tony Blair had been a historian rather than a lawyer then he might have paused momentarily before rushing in to lend British support to US military intervention, as history tells us that foreign interventions in Afghanistan tend to end in bloody disaster. The most infamous British example of this was in the first Anglo-Afghan war of 1839–42 when the British successfully carried on up the Khyber Pass, occupied Afghanistan and then withdrew the majority of their troops. Only three years later, fearing for their safety the British garrison in the capital, Kabul, made the ill-fated decision to evacuate: they numbered 4,000 soldiers and 12,000 civilians and were led by

Major-General William Elphinstone, a Scot. The Afghans had promised Elphinstone safe passage but in one of the worst disasters in the history of the Empire, as they departed Kabul the British were massacred; there was said to be only one survivor. The remaining British forces swiftly left the country having become the first Europeans to realise the painful lesson that when it came to Afghanistan it was a not a case of if occupation failed, but when occupation failed.

Robin Cook from Bellshill had been appointed foreign secretary from 1997–2001 before being, surprisingly, demoted, albeit he remained a prominent figure in Cabinet. When in March 2003 after a year of sabre-rattling and with Britain still in Afghanistan, Blair proposed that Britain now join the US in a second military invasion, this time of Iraq, Robin Cook resigned. Cook knew, more than most, that no matter what the dodgy dossiers might say there was no intelligence then available which proved, 'beyond doubt', that Saddam Hussein held weapons of mass destruction. The Americans may have been happy to say their reason for intervening in Iraq was to overthrow Saddam and change the regime but the UN Security Council and international law did not support such military intervention. Yet Robin Cook remained a lone voice in Cabinet and in the minority in the House of Commons when the decision was made to go into Iraq. The Conservatives led by Edinburgh-born Iain Duncan Smith – yes, that really did happen – supported the Government and of the three major parties, only the Liberal Democrats led by Inverness-born Charles Kennedy were firmly opposed.

For the last four years of Blair's tenure in No. 10, the war in Iraq dragged on and it became crystal clear that in the rush to invade nobody seemed to have given the slightest thought to what to do if and when Saddam was defeated. Despite Saddam's eventual capture and execution Iraq became bloodier and more brutal rather than safer and more secure, yet Blair continued to enjoy support from his government and parliament and in 2005 was re-elected. However, he was no more the bright young thing he had been when he came to power, and his permanent grin seemed more and more forced as the months went by and all those millions who on the election night of 1997 had stayed up for Portillo now began to wonder if perhaps he might not be the 'pretty straight sort of guy' he'd claimed to be.

In 2010, following the return home of British forces in 2009, former prime minister Blair was called as a witness at the Chilcott Inquiry, established to investigate Britain's role in the invasion of Iraq. Unsurprisingly Blair remained firmly of the view that despite everything that had happened the invasion of Iraq had been the right course of action, that he never misled anybody, anywhere, anytime, and that he had no regrets whatsoever. Furthermore, with his statement that we should probably invade Iran as well, it was clear that the Blair Doctrine was alive and well, no matter how many people in Iraq and Afghanistan were not.

By 2010, Blair had, to almost universal relief, been retired from frontline politics for three years. All those who had marched in 2003 to say that the war was 'Not in my name' had not changed their minds and all those who had supported the war had either changed their mind or appeared happy to let Blair take the blame for all the setbacks that had taken place after the invasion. For all his ten years in high office and his record as both Scotland's and Labour's longest-serving British prime minister, Tony Blair will have to grin and bear the uncomfortable legacy of Iraq – although considering his post-No. 10 career as arguably Scotland and Labour's wealthiest ever former prime minister he will probably get by.

Numerate Scots

In the film *Gregory's Girl*, was it not Clare Grogan who when on her date with John Gordon Sinclair felt compelled to ask, 'Why are boys so obsessed with numbers?'. And it is true that ever since the 16th century the world has associated Scotland with mathematicians, financiers and entrepreneurs, little suspecting that in 2008 Scottish bankers would nearly bankrupt the entire UK economy. But the recent travails of the Royal Bank of Scotland and the Bank of Scotland are merely a chapter in the long history of Scottish financiers who over the centuries in their eternal search for a positive balance sheet have seen many busts to go with the years of accumulated boom. Even after the near collapse of the Scottish banking system, Scots and numbers remain as inextricably linked as Oor Wullie and The Broons, mince and tatties or any other Scottish double act you care to mention – with, of course, the notable exception of Halifax and the Bank of Scotland.

LOOK AT THOSE TREES SWAY – THEY'VE GOT LOGARITHM

21 JOHN NAPIER (1550–1617)

John Napier was an Edinburgh landowner with interests in theology, military inventions and mathematics. In the dark days before the invention of the pocket calculator, working out complicated numerical sums was a much slower process. When you had numbers going up to only 1,000 or so, life was fairly straightforward, a little addition here, an occasional multiplication there, a baker's dozen to keep everyone honest, but from the 16th century onward numbers were growing bigger and bigger and it was becoming increasingly difficult to work out complicated sums. The life of clerks and accountants was onerous and overtime had become compulsory. Something had to be done and around 1594 Napier came up with the idea that rather than all that multiplying and dividing all numbers could be expressed as fractional powers of a base number, and that if you added (or subtracted) those fractions together it would be a much quicker way to come up with the final answer.

He would spend the next twenty years working out complex tables to demonstrate his system and in 1619 Napier published his findings under the catchy title *Mirifici Logarithmorum Canonis Descriptio* and it was this work that introduced logarithms to the world. The word comes from the Greek *logos*, meaning ratio, and *arithmos* meaning number, and a logarithm is described as the exponent indicating the power to which a fixed number, the base, must be raised to obtain a given number or variable – in other words, a shortcut equation to make sums easier.

Napier's logarithms revolutionised the world of mathematics and inspired English mathematician Henry Briggs to travel to Edinburgh to discuss numbers with him. Napier and Briggs agreed to adapt logarithms from natural logarithms to what would become known as common logarithms using a base of 10, and in 1617 Briggs published the first tables of logarithms for all numbers from 1 to 1000 up to eight decimal places, and over the following two decades tables were published with the logarithms for numbers up to 100,000.

In 1617 Napier published details of an abacus he had perfected to aid multiplication, division and square root calculations – an invention which proved to be the blueprint for the slide rule. The abacus comprised, on the vertical scale, squares numbered from 1 to 9 and, running horizontally, ten numbered rods from 1 to 9 with zero on the end. The rods were made of either bones, or in more upmarket versions ivory, and the abacus became known as Napier's Bones, also in commemoration of the fact that by the end of 1617 Napier was six feet under.

Ever since Napier first published his findings mathematicians have worked on logarithms to make them more accurate. In his original work Napier had talked of a base with a mathematical constant and in the 18th century the brilliant Swiss mathematician Leonard Euler worked out that this base was actually 2.71828 – or there about. Euler would call this mathematical constant 'e', and according to Euler 'e's were good.

In the 20th century logarithms to the base of 2 were used in computer science but it would be computers and pocket calculators that marked the decline of logarithms as an everyday mathematical tool. Up until the 1960s people were taught that 10 to the power of 3 using a base of 10 equalled 1000 and that the logarithm, or log for short, of 1000 to the power of 10 was therefore 3. After that point nobody had much of a clue what was going on and would often curse Napier for the many hours they had to spend studying logarithms, but thanks to slide rules and logarithm tables complex sums could be worked out and one generation of loggers would faithfully pass the book of tables on to the next. And so it continued until the 1970s and the advent of the calculator with which you could now find answers to complicated mathematical problems with the straightforward press of a key. Napier was out, Casio was in.

Logarithms may have fallen out of fashion in recent years, but for centuries they aided advances in science, engineering and astronomy by making calculations easier and more accurate, and there is one other reason why Napier should be remembered; in 1614 he became one of the first to promote the use of the decimal point rather than any of that fraction nonsense, although when you hear Scottish football managers still refer to 'a game of two halves' you realise that as far decimalisation is concerned there is still much work to be done.

LE BOOM ET LA BUST

22 JOHN LAW (1671–1729)

When John Smith became leader of the Labour Party in 1992 it was often said that his persona was that of an avuncular Scottish banker. Smith was, in fact, a Scottish advocate and not a banker, but the comment was widely perceived as a compliment reinforcing several hundred years of the Scots' reputation as sound financiers, good, and even careful, with money. Equally, you could argue that the recent RBS and HBOS debacle was not entirely out of character. For all the carefully cultivated air of Presbyterian prudence, Scots have never been risk averse, why even the dourest of Calvinists has been known to partake of a wee dram or two, and as J. M. Barrie once remarked, 'There are few more impressive sights in the world as a Scotsman on the make'.

An early example of a Scottish wheeler-dealer and a man several hundred years ahead of his time was financier and gambler John Law who was born in 1671 into an Edinburgh banking family. As a teenager he studied banking and finance but after discovering an early gift for gambling decided that the latter would be a rather more exciting way to make money. Law moved to London, but after killing a man in a duel in 1694 – whose dying words were, 'I fought the Law, and the Law won' – was found guilty of murder and fled from England. Law would spend the next twenty years in Europe gambling, speculating, writing and ingratiating himself at the French court, which finally paid off in 1715 when the French regent, the Duke of Orleans, appointed him, still a wanted felon in England, as Controller General of the Finances of France.

With plenty of time on his hands since fleeing England, Law had returned to the study of banking and finance. He came up with the radical realisation that real national wealth was not dependent on the possession of money and gold, but that trade was what made a country wealthy and that money, although very important, was a means of exchange to attain even greater levels of trade. Law wrote several pamphlets about this and various other reforms he thought pertinent. Finding himself in a position to put his ideas into practice, Law persuaded the Regent to establish the state-owned Banque General in 1716, which would become the Banque Royale in 1718. In an effort to revitalise the French economy, crippled by the five-year War of Spanish Succession that had ended in 1714, the bank would issue paper banknotes. In addition to this, Law increased the amount of credit available and transferred the national debt into shares in offshore trading companies and in particular the Mississippi Trading Company.

In 1717, Law had taken a controlling interest in a failing business called the Mississippi Company trading in the French colony of Louisiana, and the Mississippi Company and John Law were given a monopoly to trade with the French colony of Louisiana and their colonies in the West Indies. Law busily promoted the virtues of buying shares in his company alongside the benefits of trading with these colonies, and combined with the ready availability of credit and paper money the French economy was soon thriving, share prices were going through the roof and a small settlement by the name of New Orleans was founded on the Mississippi River in 1718. Life was just one long Mardi Gras.

However, as you have probably guessed, this credit bubble was not to last. John Law's shares' promotion had been so successful that in two years the Mississippi Company had managed to take on all of France's national debt and shares were now worth more than twenty times their original value, and when shareholders decided in 1720 to cash them in the company was not in a position to pay out. The system Law had created was, of course, based predominantly on confidence – most of the paper banknotes and company shares had no real money or stock backing them up. It did not take long for panic to set in, the Mississippi Company to go bankrupt and the whole financial edifice created by Law to come tumbling down. Having collapsed the economy, not even his friendship with, or any amount of ingratiation toward the Regent could save Law from being dismissed from his post and he was thus now in the unique position of being exiled from France as well as Britain.

The Banque Royale managed to survive the collapse and remained the French central bank; it was renamed the Banque de France in 1800 to fit in with the new, post-revolutionary era of liberté, egalité and frugality. As for John Law, he would return to the gambling tables elsewhere in Europe where for the remainder of his life he won some and lost some, but not with the same panache or success as he had in the past – behind his trademark tartan shades he surely reflected on a life less ordinary which had finally run out of chips.

John Law was not the first Scotsman to found a national bank. The name William Paterson is most associated with Scotland's disastrous attempt in 1698 to build itself an empire in the New World. Paterson, who was born in Dumfries and Galloway, was the driving force behind the Darien expedition. Undertaken in what is now Panama, the attempt to colonise Darien was abandoned after only two years, bankrupted the Scottish economy and was a major factor in the Act of Union seven years later. Paterson was himself one of the original colonists, but returned within a year having lost his wife and son.

However, outside Scotland Paterson's name is remembered today for a more successful financial venture. Long before Darien, Paterson had made his fortune trading in the West Indies and when he returned to London he persuaded the English government in 1694 to set up the Bank of England. Paterson would arrange a loan to the government for £1.2 million and with this money as backing, the bank would issue banknotes.

The Bank of England became the central bank of first the English and then the British government and became the model for other central banks around the world; it moved to its current location in Threadneedle Street, London in 1734. From the 1930s it became the sole source of banknotes in England and Wales and was also responsible for the country's financial reserves and managing the national debt. The bank remained a private company until 1946 when the Government nationalised it.

To describe ultimate security the saying, 'As safe as the Bank of England', became common parlance in recognition of the bank's ability to weather financial crises, however with Britain's national debt in 2010 looming larger than ever, the phrase is now going out of fashion.

THE BEST THINGS IN FIFE ARE FREE

23 ADAM SMITH (1723–90)

The Fife town of Kirkcaldy has given much more to the world than football team Raith Rovers whose victory in the Scottish League Cup Final of 1994 was so unexpected that non-Scottish commentators hitherto unfamiliar with the team reported, 'They will be dancing in the streets of Raith tonight'. Prime Minister from 2007–10, Gordon Brown [30] was born in Glasgow but he was raised in Kirkcaldy, was MP for Kirkcaldy and in 2008 he announced that he had rescued the world from economic collapse – although he neglected to mention that the UK economy had been nearly bankrupted in the process.

Before Brown became premier he held office as chancellor of the exchequer for ten years – the longest continuously serving chancellor since the 1820s. It is perhaps too soon for history to properly evaluate the legacy of the Iron Chancellor – and most historians are too busy retraining for the check-out counters at Tesco – but regardless of what may or may not be recorded on the subject of Gordon, Kirkcaldy can rest easy in the knowledge that its proud place in political economic history will remain forever through the writings of another of their favourite sons, Adam Smith.

Adam Smith made his name as a professor of philosophy at Glasgow University. He and David Hume [32] became lifelong friends and they would enjoy regular poker nights which Smith would invariably win as the ever-sceptical Hume would be unable to resist wanting to see what was in Smith's hand. The book for which Smith would become famous took him more than twenty years to write. Published in 1776, the *Wealth of Nations* was met with immediate acclaim and is universally recognised as one of the most important works in modern history and in which Smith first espoused the foundations of modern economic thinking.

Before Smith expounded his philosophy, mercantilism was all the rage with governments imposing high taxes and tariffs on trade and goods, although remaining fairly lax on health and safety, and maternity leave could last up to twenty years, unpaid.

Smith proposed that governments would keep out of commerce as much as possible and let individuals have free reign. He did not use the term 'capitalism' but did describe 'capital' as stock and profits and that it was a goal of economic and political systems to improve that capital by removing state restrictions and promoting free trade between countries. To help achieve this he suggested a 'division of labour' so that workers would be employed in roles that suited them best for the benefit of all. Smith believed that economies would prosper through competition and put forward the concept of the 'invisible hand' as a description of goods and services being allowed to find their true value in the market place; to be guided, as it were, by the invisible hand. Adam Smith became known as the Father of Modern Economics and with the Industrial Revolution just around the corner his ideas received an appreciative response from businessmen, merchants and factory owners throughout the western world who were all more than happy at the idea of paying less tax.

Throughout the 19th century the *Wealth of Nations* would influence politicians, economists and philosophers worldwide. Smith became associated with supporting the virtues of laissez-faire (in terms of the economy), although in fact he never supported unrestricted commerce between countries, and laissez-faire was taken up by Victorian imperialists who got around the problem of national borders by the straightforward stratagem of conquering a quarter of the world. At around the same time, Smith's proposals on the division of labour and the requirement of society to look after its less fortunate members would also inspire Karl Marx in *Das Kapital*, although Marx ended up with somewhat different conclusions. And we should not forget Smith's influence on that other Marx, Groucho, who once said that, 'the secret of life is honesty and fair dealing – and if you can fake that, you've got it made'.

Adam Smith fell out of favour in the 20th century with the rise of Communism and Marxism and the advent of increased state intervention as proposed by the influential English economist John Maynard Keynes, but with Reaganomics and Thatcherism in the 1980s Smith came back into vogue when his name was re-appropriated as an advocate for right-wing, free-market economics – although we cannot blame Smith for Reagan and Thatcher changing the 'for the benefit of all' part of his analysis to, 'for the benefit of the forty percent who will re-elect us'.

It was the 19th-century, Dumfriesshire-born writer and historian Thomas Carlyle who once declared that economics was a 'dismal science', although if Carlyle were around today even he might have himself tuning in to BBC News business editor Robert Peston telling us which particular financial institution had managed to go bust that day, with ALL the emphasis ON all THE wrong words. As we enter the 2010s it turns out that the free market is actually not free at all, but will cost us all trillions of pounds, dollars and euros, with the division of labour being replaced by the division of debt. But at least we now know that Adam Smith has been proved right about one thing, that the 'invisible hand' that guides the free market and has built in checks and balances is, as Smith said, invisible. Dismal days indeed.

FOR ALL THE DRUGS IN CHINA

24 WILLIAM JARDINE (1784–1843)

Amid all the talk of smog, Usain Bolt and whether Tibetan monk-throwing was a legitimate world sport, one fact that was overlooked during the 2008 Beijing Olympics was the story of the first ever Chinese Olympic gold medallist. Li Airui was born in the north-eastern city of Tianjin and left China at the age of six when his family moved to England. But after studying for four years at Edinburgh University he won the 400-yards at the Paris Olympics in 1924, returned to Tianjin and remained in China for the rest of his life, eventually dying in a Japanese internment camp in 1945. In the English-speaking world Li Airui is better known as Eric Liddell. The son of Scottish Protestant missionaries, he himself would become a missionary in China and his athletic exploits are recounted in the Oscar-winning film *Chariots of Fire*, in which he famously declared that because of his religious beliefs he would not run on the Sabbath, or at least not until he had first read the *Sunday Post* and sadly there were no copies to be found anywhere in France.

Missionaries were not the only Scots in China. In 1832 in Canton (now Guangzhou) William Jardine, a surgeon-turned-merchant born near Lockerbie, founded with another Scottish merchant, James Matheson from Sutherland, a trading company that then became established in southeast China by the name of Jardine Matheson, or Jardines for short. The British were by this stage well into their national obsession with tea drinking and Jardine Matheson exported tea from China and in return supplied the Chinese with their own national obsession, opium, which was imported from British India. Jardine and Matheson became so successful that they were among the first ever foreigners to gain the honorary title, *tai-pan*, a Cantonese term for business bosses meaning, literally, 'top class'. Jardine also gained the nickname, 'iron-headed old rat', after he was struck on the head by an iron bell that fell from a scaffold, and yet continued to walk on as if nothing had happened. The Qing emperors of China were less enamoured with these Scottish barbarians, and with the millions of Chinese drug addicts, and were determined to end their countrymen's love of the pipe so, in 1839 they banned the import of opium.

Jardine returned to Britain and set about persuading the government that the rights of Chinese opium smokers should be defended at all costs and that the Chinese authorities should be forced to reverse their decision – and kept passing a bong around until Ministers agreed with him. This led to the First Opium War of 1841 and the British occupation of a small island by the name of Hong Kong, meaning 'fragrant harbour' in Cantonese, which Jardine had coincidentally pointed out would be ideal as a strategic trading port. The following year the Chinese capitulated to British demands and with the Treaty of Nanking (or Nanjing) allowed the continuation of the opium trade (although diplomatically opium was never mentioned in any of the treaty documents), the opening up of Chinese ports including Canton and Shanghai to international trade and ceded Hong Kong to the British. The acquisition of Hong Kong amused the generally not amused Queen Victoria, who was somewhat surprised to find herself queen of a small, insignificant island with a funny name, although she had to admit that the sweet and sour spare ribs were perfectly acceptable.

Jardine Matheson were quick to buy up the first plots of land and set up base in Hong Kong and were at the forefront of the colony's growth throughout the 19th century. Jardines also expanded into mainland China where, for a period before World War II, their head office was based in Shanghai, and when Japan opened up international trade in the 1850s Jardines were the first foreign company to buy land in the new port of Yokohama in 1858. This move into Japan was led by Borders-born William Keswick, a great-nephew of William Jardine, and the Keswick family have retained control of the company to this day.

Today, despite World War II, China turning Communist and Hong Kong being returned to China in 1997, Jardine Matheson Holdings – which has expanded into construction, property, retail, finance and hospitality with numerous company names beginning with Jardine – remains the largest Asian-based multinational and the largest commercial company in Hong Kong. Hong Kong itself has become one of the globe's major financial centres and is home to the world's largest bank, the Hong Kong and Shanghai Banking Corporation (HSBC), based there since its foundation in 1865 by Thomas Sutherland [25] from Aberdeen.

The opium trade which had helped make Jardines went into decline at the end of the 19th century when the Chinese authorities decided to follow the maxim, 'If you can't beat them join them', and began to grow their own opium, until, finally, Chairman Mao oversaw the trade's eradication in China with a carefully thought-out public health strategy of death to the opium farmers.

As for the two original Scottish *tai-pans*, William Jardine and James Matheson, both returned to Britain extremely rich. Matheson left China in 1842, bought the entire island of Lewis in 1844, became a Member of Parliament and died in France in 1878. Jardine, although continuing as head of the company, had remained in Britain after 1839, and was also elected a MP in 1841. William Jardine died in 1843, one of the most powerful and wealthiest men in the country, and his funeral was attended by many prominent figures, who one would like to think marked the event by getting completely wasted.

Hong Kong is not the only major financial centre with a Scottish connection. The Englishman Stamford Raffles founded the city-state of Singapore in 1819, but although Raffles is forever associated with the island he only stayed one night in the new settlement, and the task of turning Singapore into a strategically important and thriving colonial port was handed over to his deputy, William Farquhar from Aberdeenshire. Farquhar was appointed Britain's first resident in 1819 and when he was replaced in 1823 it was by another Scot, John Crawford from the island of Islay. With the exception of World War II, Singapore would remain under British rule until 1963 and although it gained its independence in that year English remains the principal language.

Today's Singapore is one of the wealthiest, most expensive and, thanks to their strict anti-littering laws, one of the cleanest places in the world – and the canny Scots Farquhar and Crawford would no doubt approve heartily of any scheme where you charge 1,000 dollars for dropping a sweetie wrapper, although as a native of Islay and the home of some of the finest of malt whiskies Crawford would never ever have condescended to partaking of something as God-forsaken as a Singapore Sling.

THE HISTORY OF SCOTTISH BANKING (INCLUDES BONUS FEATURES)

25 THOMAS SUTHERLAND (1834–1922)

William Paterson may have founded the Bank of England but he later, through the Darien expedition of 1698–1700, bankrupted Scotland to such an extent that within seven years it ceased to be an independent nation. And John Law [22] may have introduced one of Europe's first credit booms when he promoted paper money and shares in France in 1716, but this only resulted in the collapse of the French economy four years later. So, when it comes to financiers reaching for the stars then crash-landing, the Scots do have form.

The Royal Bank of Scotland (RBS) took more than 250 years to become an international success. Founded in Edinburgh in 1727 as a nouveau riche upstart to the Bank of Scotland established some thirty years earlier, one of its initial attempts to gain business was to give what is thought to be the world's very first overdrafts when it offered Edinburgh merchant William Hogg a £1,000 credit limit, although in a sign of things to come it is not actually recorded if Hogg ever paid the money back.

Up until 1751 the two banks did their best to destroy each other before realising that if the upper and middle classes, farmers, Conservative voters and anybody who aspired to hosting a dinner party all joined the RBS and everybody else joined the Bank of Scotland, then there would be more than enough business to keep both of them happy as any Scottish banker can ever be. This equilibrium continued until the 1980s when RBS, who had acquired several small English banks since the 1920s, began to trade in England and, having dipped their toe in international waters by opening their first overseas branch in New York in 1960, bought the American Citizens Financial Group in 1988. The next step up for RBS came in 2000 when they defeated the Bank of Scotland in the race to take over NatWest – an English bank three times their size – which placed RBS firmly in the big league.

By 2007 Royal Bank of Scotland was increasingly being re-branded as RBS, although whether it was the Royal part, the Scotland part, or both, that it was embarrassed by was not made clear. 2007 was also the year that RBS under Paisley-born Fred Goodwin led a £49 billion consortium to buy out the Dutch banking group ABN AMRO, and in the process made itself the second largest banking group in Europe, the eighth largest banking group in the US and the tenth largest company in the world.

By this stage RBS was making annual profits of more

than £10 billion and employing more than 170,000 people worldwide; it owned Ulster Bank, Churchill Insurance, Direct Line Insurance and sponsored Formula 1, the Rugby Union Six Nations and as far as first time visitors to Edinburgh Airport were concerned, seemed to be the official sponsor for the whole of Scotland.

Two years later, RBS announced, in respect of its £24 billion loss – the largest annual loss in British corporate history and only £4 billion less than the entire Scottish government budget for 2009 – was eighty percent owned by the UK Government which had been forced to pump in billions of pounds to keep the bank afloat. All of which only goes to show that all those Scots who had been suspicious of RBS for the past 250 years were right all along – you should never trust any financial organisation that has Royal in its name.

Scots have been ever keen to found banks the world over and while Scottish banking has recently nearly imploded both Australia and Canada have seen considerable international growth. In Australia it was the Scots-born governor-general of New South Wales, Lachlan Macquarie [43], who founded Australia's first bank in 1817, and it was Scots-born Australian prime minister Andrew Fisher [50] who established what is now Australia's largest bank, the Commonwealth Bank, in 1911. The Standard Chartered Bank, one of the largest banks in Asia and Africa, was founded in 1969 and is the merger of two banks, the Chartered Bank of India, Australia and China, founded in 1853 by James Wilson from Hawick, and the Standard Bank of South Africa, founded by Scotsman John Patterson in 1862. Scottish merchants founded the oldest bank in Canada, the Bank of Montreal, in 1817, and fifteen years later in 1832 the Bank of Nova Scotia, later to become Scotiabank, was established in Halifax, Nova Scotia, which just goes to show that it is not always a complete disaster when a bank with a Scottish name and Halifax go together.

When RBS was struggling through 2008, Barclay's replaced them as the second largest bank in the UK, with Barclay's taking its name from James Barclay; the London-born son of Scottish Quakers from Aberdeenshire. However the number one bank in Britain, Europe and for that matter the world in both 2008 and the subsequent years remains the Hong Kong and Shanghai Banking Corporation (HSBC), founded in 1864 in Hong Kong by an Aberdonian, Thomas Sutherland. It would come as a surprise to Sutherland to find himself best remembered for this, as during his lifetime he was most renowned for his long career with the Peninsular and Oriental Steam Navigation Company, or as it is better known, P&O.

Arthur Anderson from Shetland rose from shipping

clerk to co-found the original Peninsular Steam Navigation Company in 1835 that would take its longer moniker in 1840 and made its reputation by carrying mail and passengers to the ports of Spain and Portugal – hence the flags featured on the company logo – and by the mid-19th century owned the largest fleet of commercial steamships in the world.

Thomas Sutherland, also formerly a shipping clerk, joined P&O in 1852 aged eighteen and made his name in Hong Kong where he was promoted to manager. In 1872 he was appointed managing director and would further emulate Anderson by becoming chairman of P&O in 1881 and a MP in '84. Sutherland was chairman of the shipping line until his retirement in 1914 having been responsible for the safe passage of passengers and parcels from Britain to the three Empire mainstays – India, the Far East and Australia – for forty-two years and having worked for the line for more than sixty years, which presumably meant that his retirement present must have been a very large clock indeed.

While Sutherland was managing P&O's Hong Kong operation he noticed the lack of banking facilities and opened his first branch in 1865 at 1 Queen's Road, Central, which remains HSBC's headquarters today; a branch in the other major European trading port, Shanghai, followed one month later. HSBC enjoyed steady growth in China and Asia up until the 1950s, when through a series of acquisitions and expansion into new markets it began its rise to become the world's largest banking institution. And all because an Aberdonian wanted somewhere safe to leave his wallet . . .

FIFE AID

26 ANDREW CARNEGIE (1835-1919)

The history of the modern steel industry began in 1855 when English engineer Henry Bessemer patented his process to mass-produce steel from molten pig iron. Britain was initially at the forefront, producing nearly half the world's steel up until the 1870s with Scotland producing a disproportionate twenty percent. Up until the end of the century production in Britain, and especially Scotland, continued to grow in line with demand at home, but in global terms Britain fell to producing just under twenty-five percent – production in other countries had caught up and one in particular, the US, became far and away the dominant steel producer. The US would, as a result, also become the world's dominant industrial nation. And at the forefront of America's ascendancy was wee a guy from Fife, the original 'man of steel'.

Andrew Carnegie was born in Dunfermline and his

family immigrated to America in 1848 and settled in Allegheny, Pennsylvania, near the city of Pittsburgh that was founded in 1758 by a British general, John Forbes who, coincidentally, also hailed from Dunfermline. Forbes had named the settlement in honour of the British politician William Pitt the Elder and by adding the suffix, 'burgh', he had the wildly optimistic hope that Americans might learn how to properly pronounce Edinburgh.

Carnegie began working in a cotton mill at the age of thirteen before becoming a messenger boy for a telegraph company. When he was eighteen he joined the Pennsylvania Railway Company and over the next dozen years he progressed through the company and began to make financial investments in railways and oil that would bring him great dividends. In 1865 he set up his own ironworks, followed by a steelworks in Pittsburgh in the 1870s and over the next thirty years, through astute and often ruthless management, he built up an empire producing steel for the railway lines criss-crossing the American continent.

By the end of the century Carnegie was producing a quarter of the US's total steel output. The Carnegie Steel Company was the largest industrial company in the world and calculations suggest that taking into account inflation Carnegie became one of the richest men the world has ever known and unquestionably the Fifer most able to stand a round – although being a Fifer he'd want the change. In 1901 Carnegie decided to retire from the steel industry and dedicate himself to a life of philanthropy, a career change eased by his $300 million personal cut when he sold his company.

Andrew Carnegie was, for the most part, self-educated and as a young man had spent many hours in his local Pittsburgh library immersed in books. He stated that the first third of life should be dedicated to educating yourself, the second third to making as much money as possible and the remaining third to giving away as much money as possible to worthwhile causes. Carnegie backed this up by funding more than 2,800 libraries around the world that would be free to the public, as well as making donations to numerous universities, museums, scientific institutes and educational trusts. In total, in his lifetime Andrew Carnegie gave away more than $380 million to good causes, and is possibly the greatest philanthropist the world has ever known. Carnegie also stated that no man needed more than $50,000 a year to live on, which was still about the same as the entire population of Dunfermline at the time, but did mean that he was definitely not the man to have as your phone a friend when the hundred thousands dollar question came up in *Who Wants To Be A Millionaire*.

Today the philanthropist's name is most universally recognised for the Carnegie Mellon University in Pittsburgh and Carnegie Hall in New York. The university is one of the most prestigious in the world and began life in 1900 as the Carnegie Technical School until in 1967 the Carnegie board persuaded a representative of the Mellon Institute of Industrial Research that the two institutions should merge – an early example of, 'twisting the mellon man'. Carnegie Hall, New York, as opposed to Carnegie Hall, Dunfermline, was built in 1891 on Seventh Avenue, Manhattan and is one of the most famous and finest concert halls in the US and gave credence to the well-known showbiz saying:

Question: 'Pardon me, sir, how do I get to Carnegie Hall?'

Answer: 'Practice, practice, practice.'

The same question is also well known in Dunfermline where the answer is, 'Up the High Street, next to the roundabout'.

While he spent most of his life in America, Carnegie never forgot his Scottish roots and made many endowments and donations to Scottish institutions, with one of the most personal involving Dunfermline's Pittencrieff Park, known locally as The Glen. When he was a boy Carnegie would often go to the gates of the park, but it was privately owned and he was never allowed in. Having made his fortune he bought the Park in 1902 and donated it to the people of Dunfermline so that no child need be excluded as he had been – much to the exasperation of generations of parents since who had no idea where their children were.

OLD MACDONALD HAD A WIND FARM, EH-OH, EH-OH, PO

27 JAMES BLYTH (1839–1906)

Those who are the not the greatest fans of Alex Salmond, and there must be a few, might smile an ironic smile if the First Minister, who is often described as being even more long-winded and full of hot air than your average long-winded and full of hot air politician, were to go down in history as the politician who oversaw a new golden era of Scottish economic prosperity through the growth of wind power. For the many who thought renewable energy meant no more than a few Teletubbies' windmills in the back of beyond, and that was all very well but would never amount to anything, and you wouldn't want them in your back garden even if your three-year-old thought they were cool, it was the opening in 2009 of Whitelee Wind Farm, only nine miles from Glasgow – the then largest farm in Europe with 140 turbines – that made them think there might be more to this Green racket than remembering not to leave the telly on standby all night.

Man has been utilising wind power for thousands of years with the first windmills built around the 7th century in Asia and in Europe during the 12th century. Wind power was first used to grind corn and pump water, but with the invention of James Watt's [52] steam engine in the 18th century, windmills fell into decline with their only purpose remaining as subjects for art students too shy to paint naked women. Interest in wind power only revived in the wake of oil shortages during the 1970s and various European countries and some States in the US began spending time and money researching alternative forms of energy.

Overall, progress has been slow and by the end of 2009 only 1.5 percent of the world's electricity was derived from wind power, although that figure was at least four times what it had been in the previous decade and is set to increase substantially in the decade to come. At the time of writing, Scotland has the capacity of producing just over 1 percent of the world's total wind power and with more on- and off-shore ventures scheduled for the future it is possible that – given continued investment, sustainable sites, political will and, if you pardon the expression, a fair wind – Scotland has a fighting chance of achieving its ambitious target:

fifty percent of the country's electricity requirements to be supplied from renewables by 2020 – unless of course the experts change their minds yet again and decide to go all nuclear on us.

If Scotland can stake a claim at the forefront of green energy-production then it is only appropriate because it was a Scot in the 19th century who was the first person to produce electricity from wind power.

The first electricity-producing windmill is often said to be that of American inventor Charles F. Brush who toward the end of 1887 built a massive, 60-foot (18 m) turbine with 144 rotor blades that charged the batteries that powered his home in Cleveland, Ohio. Brush's wind turbine was very impressive and much admired, but unbeknown to him, James Blyth, an academic from Aberdeenshire, had beaten him to it by four months.

Blyth was born in the village of Marykirk on the Aberdeenshire/Angus border. He graduated from Edinburgh University, taught mathematics and in 1880 was appointed professor at what is now Stratchlyde University in Glasgow. Blyth retained a home in Marykirk where he would return for holidays and rather than re-charging his batteries by going fishing or taking long walks he preferred to take some batteries to Marykirk and come up with new ways of re-charging them.

Blyth had long been interested in the possibility of

using wind power to create electricity and in July 1887 was able to successfully light up his Marykirk home using power from a thirty-foot, cloth-sailed turbine. Although it worked, Blyth was not satisfied and built a further two, more efficient models which would illuminate his holiday home until the end of the century and for which he received a UK patent in 1891. Blyth's turbines produced more than enough power for his needs and he offered the excess to the village, but was turned down by his neighbours who considered electricity deeply suspicious and possibly the work of the devil. However, some eight years later he would have more success persuading the lunatic asylum at nearby Montrose to try one, as many of the residents thought there was a lot to be said for the work of the devil.

After Blyth's death in 1906, his windmills were dismantled and abandoned and his own country neglected the technology he had pioneered. It would be other European countries such as Denmark that would be at the forefront of developing wind power further and it would be the best part of a century before Scotland began to catch up. With Scotland now setting itself the target to become a net exporter of renewable energy in the coming decades one can look forward to the time when the old nationalist slogan, 'It's Scotland's Oil', will become the all-new, 'It's Scotland's Wind' –

although as many long-suffering Scottish wives and girlfriends can testify to, an excess of Scottish wind is not necessarily something to be proud of.

YOU'LL HAVE HAD YOUR TEA

28 THOMAS LIPTON (1850–1931)

Sri Lanka is beautiful – from stunning beaches to elephant orphanages and from sacred Buddhist temples to ancient ruined cities – just 'don't mention the war'. Kandy is a bustling metropolis, well worth a visit, as is the capital, Colombo, and if you ever happen to be in the latter you may be surprised to come upon a large department store by the name of Cargills, and you may be even more surprised to find out that this department store was founded back in 1844 by two Scots, David Sime Cargill and William Milne. The success of the business would lead to the Angus-born Cargill founding, in 1886, Burmah Oil in Glasgow – Britain's first international oil company which made its money drilling in South East Asia and remained a major player until it was bought by BP in 2000.

There was a certain irony concerning BP's acquisition. Oil had been discovered in Iran in 1908 and the following year when BP's forerunner, the Anglo-Persian Oil Company, was founded to capitalise on the find it was Burmah Oil that became both the major investor and shareholder in the new company. In 1913 John Cargill, son of David Sime Cargill, and the new company chairman, decided to sell Burmah's shares to the British government. With hindsight this could be said not to have been one of the greatest business decisions of all time, but when your predominant location of business is Burma then do you really want Iran to be your second?

Sri Lanka was ruled by the British from 1796 until it gained independence in 1948 and during that period was known as Ceylon and became famous as one of the largest tea producers in the world. It was the Dutch who first brought tea from China and Japan to Europe in the early 17th century, where it was initially a drink that only the aristocracy could afford. There is no record of tea in England and Scotland until the end of the 17th century when it was noted that, due to its high cost tea had to be carefully conserved – so beginning the longstanding Scottish belief that the additional 'one for the pot' is extravagant.

By the end of the 18th century the price of tea had dropped and Britain's obsession with the cuppa as the antidote to all physical and psychological problems had become established throughout the land. Tea in Britain was imported from China by the East India Company and to ensure that it arrived as speedily as possible, fast sailing ships called clippers were built. The most famous of the tea clippers was the *Cutty Sark*, built in Dumbarton in 1869 and named after the attractive witch from Robert Burns's [34] *Tam o' Shanter*, although tea was probably the last thing on the poet's mind when he came up with the scantily-dressed character.

Realising how deeply reliant they were on China's supply, the British began to look for alternative sources of their new-found favourite drink. In 1823 a Scot called Robert Bruce (no, not that one) was told about a wild plant growing in Assam in North East India that resembled tea. Bruce sampled it and found that, although different from Chinese tea, this Assam black tea did indeed taste good. Robert Bruce died the following year; Assam tea was not a contributing factor, and his brother Charles cultivated the first consignments in 1836.

Assam was not the only tea-producing region in India. Robert Fortune a Scottish botanist from the Borders went on two surreptitious journeys to China in the 1840s – the first in search of rare plants, the second to steal as many of these rare plants as possible. As the purpose of his journey was both illegal and dangerous Fortune eschewed the traditional modus operandi of talking as loudly as possible, and in an effort to blend in shaved his head, dressed in local clothes, learned basic Mandarin and took a Chinese name, so that before long Chow

Mein, the tall, bald stranger with the weird accent was a regular attendee at garden centres all over China. Fortune smuggled more than 20,000 samples of tea out of the country and these were planted in Darjeeling and various other parts of northern India. By the end of the century, Fortune's stolen plants and combined tea from with Assam had allowed India to outstrip China and become, as she is today, the largest tea-grower in the world; Mother India now produces more than 900,000 tonnes of tea a year. Thanks to the Bruce brothers and Robert Fortune the British could finally rest easy, safe in the knowledge that there would always be time for tea.

The man who above all would become synonymous with tea was born in Glasgow and his parents ran a shop in the Gorbals. As a teenager Thomas Lipton travelled around America, working in various jobs including that of a shop assistant in New York, before returning to Glasgow and setting up his first grocer's shop in 1871. Lipton's retail empire soon grew, partly thanks to the advertising and marketing techniques he had learned in America, and by 1890 he had more than 300 stores throughout Britain and, with the price of tea falling and demand growing, Lipton decided to enter the tea market.

He looked not to India, but farther south to what was then Ceylon where another Scot, James Taylor from Aberdeenshire, had founded the first tea plantation at

Loolecondera in Kandy in 1867. In the 1890s Lipton bought five plantations in Ceylon and began to supply direct, first to his own stores and then anybody else who, as his strap-line promised, wanted tea, 'direct from the tea gardens to the tea pot'. Aimed at the mass market, Lipton's tea was packaged in tins and packets to keep it fresh and Lipton Yellow Label was sold around the globe. Although Lipton supermarkets disappeared in the 1980s, Lipton Tea remains one of the most popular brands in the world, although for some reason you still cannot get a decent cuppa anywhere in the US.

Since the country gained independence the name Ceylon has been retained only for tea-branding purposes and, today, Sri Lanka is the fourth largest tea-producer in the world and its second biggest exporter with more than 200,000 Sri Lankans working on the plantations. James Taylor, who began it all, may not have gained the international recognition that his fellow Scot Lipton achieved, but he did have the consolation of producing some highly regarded folk-rock albums.

GREED IS GUID

29 BERTIE C. FORBES (1880–1954)

Aberdeen-born comedy writer and performer Graeme Garden came up with the idea for a radio programme subtitled the 'antidote to panel games', first broadcast in 1972 and, ever since, millions around the world have tuned in to BBC Radio 4's *I'm Sorry I Haven't a Clue*, the silliest show ever invented in which, as devotees are quick to tell you, points mean prizes. Part of the success of the programme has been the use of running jokes, funny when they were first aired and still funny nearly forty years on. And one punch line featured almost invariably is, 'Gordon Bennett!'. But who was Gordon Bennett?

There were in fact two Gordon Bennetts, father (a Scot) and son (born in the US), and they both owned and ran the *New York Herald* newspaper in the 19th century. The exclamation 'Gordon Bennett!' is believed to have been coined in honour of James Gordon Bennett Jnr, whose flamboyant and often erratic playboy lifestyle led to lots of, 'Gordon Bennett! You'll never guess what he's gone and done now'. Tellingly most of Gordon Bennett Jnr's excesses took place after his father had died in 1872, although James Gordon Bennett Snr had been even more unconventional than his son.

Gordon Bennett Snr was born near Keith in Moray and initially planned to join the priesthood before abandoning that idea and emigrating in 1820, first to Nova Scotia and then to the US where, in New York, he found his true vocation as a journalist. In 1835 he founded the *New York Herald* and set out to revolutionise the newspaper industry. The newspaper was sold on the stands for one cent and for the many who were appalled to see stories of murder and scandal on the front cover there were many more who wanted more and soon the paper had the largest circulation in the US.

Bennett Snr pioneered contemporary newspaper journalism – the *New York Herald* was the first to introduce the newspaper interview, and first to pay for the privilege. He was also idiosyncratically independent in his political outlook, following no one political party or viewpoint, but through his paper's high readership remaining extremely influential. Bennett crucially supported the Union side in the Civil War, but persistently criticised Abraham Lincoln, until his assassination when the *New York Herald* was instrumental in turning the late President into an American icon.

The *New York Herald* ceased publishing as a daily in 1924, but the name lives on in the world's most popular international newspaper, the *International Herald Tribune*, throughout the globe the English language paper of last resort, always in the vain hope that it might have the Scottish League Cup results.

James Gordon Bennett Snr was not the only journalist from the north-east of Scotland to make a lasting impression in America. Bertie [Robert] Charles Forbes was born in the Aberdeenshire village of New Deer and studied at Dundee University before emigrating, via South Africa, to the US in 1904. He made his name in New York as a financial reporter and in 1917 founded the business magazine, *Forbes*, which he ran until his death in 1954 and which is owned by the Forbes family to this day. The magazine was aimed unashamedly at the movers, shakers and uptown stirrers of Wall Street and has remained consistently upbeat about the virtues of American capitalism. With Bertie Forbes stating that, 'Business was originated to produce happiness', and the Forbes family becoming one of the wealthiest in America, Bertie was obviously happier than most.

Today, *Forbes* is best known for its series of popular annual reports such as the *Forbes 400* richest Americans and the *Forbes 500s* biggest American companies, with 'Gordon Bennett!' being one of the milder oaths expressed by multi-millionaires when they read that they have lost $50 million in the past year at the same time as their accountant has bought a couple of yachts. One suspects that the great Scottish political economist Adam Smith [23] would not be too impressed by this very un-Scottish celebration of excessive wealth with which the name Forbes has become synonymous. But in one regard the family have remained true to their Scottish roots as in all the lists of the select few who one day might own a English premier league football club the name of Forbes itself is always conspicuously absent from any list of wealthiest people. After all, as any Scot will tell you, it is one thing to have lots of money, but much like owning more than one wallet, actually talking about is just plain showing off.

TEXTURE LIKE SUN

30 GORDON BROWN (1951–)

At the height of the global financial crisis in 2008 there was a heart-stopping moment when it seemed that all the 1.5 million Automated Teller Machines (ATM) in the world might simultaneously run out of money. It had been enough of a shock to realise that many of the world's largest banking institutions were broke – and furthermore were broke several times over – and even worse was the terrible realisation that property prices did not always go up, that food prices did not always go down and that sometimes plane fares cost more than £4.99, yet this was as nothing to the apocalyptic vision of a world where there was no money in the cash machine. So dependent are we on the hole-in-the-wall that, along with mobile phones, personal computers and circular tea bags, we cannot imagine life them. Yet the ATM was invented only some forty years ago and it was a Scot from Paisley who played a crucial role in coming up with the radical innovation of being able to access your account without having some twelve-year-old trying to sell you home insurance.

The first modern ATM in the world is generally recognised as the one that was opened in London in 1967 and had been developed by John Shepherd-Barron who was born in India to a Scottish family. Shepherd-Barron had come up with a special type of cheque that had a unique, mildly radioactive encryption; you inserted the cheque in the ATM, input a four-digit Personal Identification Number (or PIN) and the machine dispensed banknotes.

At the same time as Shepherd-Barron was developing his system, a Paisley-born engineer and instrument maker called James Goodfellow was working on an alternative version and a year before the Shepherd-Barron ATM was launched, Goodfellow received a patent for his machine. Goodfellow's device also required a PIN number to activate it but, instead of cheques, his used a specially encrypted card. The first Goodfellow ATM was developed a month after Shepherd-Barron's, also in 1967, and although there had been previous attempts at creating an ATM dating all the way back to the 1930s, it was Goodfellow's card system ATM rather than Shepherd-Barron's cheque system that has become the world standard. A common factor in both designs was the use of a four-digit personal PIN number as both men had realised that most people would not be able to remember a code greater than four numbers, and Goodfellow knew that for the good people of Paisley even four might be a bit of a stretch.

During 2008 it became increasingly obvious that a great many countries around the world had been enjoying unprecedented economic growth based on what appeared to be a fiscal policy of double or quits. Globally, it was a chastening time for the political and financial elite and particularly fraught for the British Prime Minister, in office since 2007, whose reputation was based on his record for successfully running the British economy for more than a decade during his tenure as Chancellor.

Gordon Brown – born in Glasgow, raised in Kirkcaldy, Fife – first came to prominence in 1972 when, while studying for a PhD in history at Edinburgh, he was elected Rector of the university. He was elected MP for Dunfermline in 1983, joined the Shadow Cabinet in 1987 and was appointed Shadow Chancellor in 1992. From 1997–2007 the New Labour alliance between Tony Blair (20) and Brown became the Ant and Dec of British politics – omnipotent and omnipresent even if, unlike the real Ant and Dec, you could tell them apart and they didn't seem to like each other very much. There were, however, areas where Blair and Brown did undoubtedly share loyalty, such as in their aim to make New Labour the natural party of government and to deliver centrist, pro-market economic policy combined with large capital investment in state provided services. They were rewarded with Labour being comfortably re-elected in 2001 and 2005 as the economy (for most) continued to flourish, and the sceptics who queried the growing debt and the growing public expenditure were recommended to lighten up and take out yet another credit card.

During the early years of New Labour's new administration, Blair came across as the shining star with Brown the dour, Iron Chancellor who was never cool and didn't holiday in Tuscan villas or the Caribbean. But when Tony's dreams and ambitions foundered in the sands of Iraq, the Blair–Brown balance shifted. Blair was forced to announce in 2004 that he would not go on and on and on as prime minister and after sidelining his chancellor at the 2005 General Election had to reluctantly bring Brown back in to shore up Labour support. Once again, Gordon was the heir apparent and, with the Arctic Monkeys playing on his iPod, he was in no mood to allow another chance to pass him by. In June 2007 after realising that he could no longer deny the clunking fist of the immovable object that was

Brown, Blair finally accepted that the force was no longer with him and the promise he made all those years ago over cappuccinos at Granita would have to be honoured. Gordon Brown became the seventh Scottish-born Prime Minister of the United Kingdom. For all those who had become completely disillusioned with the sixth Scottish-born Prime Minister of the United Kingdom things could, finally, only get better, but in fact things were to become far, far worse.

Gordon Brown's premiership turned out to be one of economic crisis, unprecedented levels of national debt, the most corrupt parliament in generations, unending conflict in Afghanistan and the end of the Harry Potter books, although for all those socialists who had never lost faith in the original spirit of Keir Hardie [10], Brown's residence in No.10 will always be fondly remembered as the time that the banks were finally nationalised.

Considering their history you do wonder why Scottish-born politicians remain so keen to have the top job. The Earl of Bute was forced out after less than year, the Earl Of Aberdeen [14] had to resign over the Crimean War, Arthur Balfour [15] was voted out in one of the largest electoral defeats in British political history, Henry Campbell-Bannerman [16] succumbed to ill-health after only two years in office, Ramsay

Macdonald [18] stayed in power only by splitting his party and became a byword for political betrayal, and Brown lost the only General Election that he ever fought as leader with even the Lib Dems unwilling to bail him out, with only Tony Blair resigning office on his own terms after a long tenure of stable government, but his legacy was such that nobody mourned his departure

The blighted Brown government was the final chapter of the New Labour project leaving a legacy of the UK's public finances being crippled for years to come as a result of the bail-out for failed Scottish banks. Or to put it another way, Scotland's final revenge for Margaret Thatcher, North Sea Oil, the Poll Tax and the 1966 World Cup Final. Perhaps the end of Gordon Brown's premiership will mark the last time that a Scot will ever be given the keys to No. 10 and the English will finally have the self-confidence to run their own affairs for a change.

Literary Scots

So what has Presbyterianism ever given Scotland other than stubbornness, the Boys' Brigade, Glasgow Rangers, a Calvinist work ethic and acute self-confidence issues? Well in the words of a certain Edinburgh-born former prime minister it gave us, 'Education, education, education'. Through hard work and learning you could improve and better yourself, no matter what background you came from – 'we're a' Jock Tamson's bairns', after all. Why, even women would be able to learn something useful – although it usually ended up being needlework.

HISTORIC CUTLERY SET FOR SALE, ROYAL CONNECTIONS AND ABSOLUTELY SPOTLESS

31 MACBETH (d.1057)

There was Mac Bethad mac Findláich, second king of the unified nation of Alba – Gaelic for Scotland – dead now for more than 500 years and relatively satisfied with his minor place in Scottish history. Alba – Gaelic for Scotland – came into existence around 1034 when the ancient Kingdom of Strathclyde was added to those of the Picts, the Scots and the Angles of Lothian. Granted Mac Bethad mac Findláich's had ended violently when the King was killed in battle in 1057 by the forces of the future Malcolm III, but a lot of monarchs came to a sticky end back then and Mac Bethad mac Findláich – Gaelic for Macbeth – at least had the satisfaction of knowing that his reign had been a period of general prosperity and stability – and to such an extent that in 1050 Macbeth had felt it safe enough to travel with his queen on a pilgrimage to Rome. There, they took in all the sights and his wife showed early signs of obsessive cleanliness by insisting that she wash her hands in the Trevi Fountain every day.

And then in 1603 everything changed. 1603 saw the death of Elizabeth I, the Virgin Queen of England, and the accession to the English throne of her nearest relative, first cousin once removed, James VI [11], King of Scotland, as opposed to James's mother, Mary, Queen Of Scots [3], whose head her first cousin had already removed. To commemorate a Scot becoming king of England, the world's greatest playwright of all time, William Shakespeare, decided to pen one of his famous theatrical entertainments and to give it a Scottish theme. Unfortunately Shakespeare was going through one of his melancholy periods – *Othello* and *King Lear* were being staged around the same time – and his Scottish Play was never likely to have lots of jokes in it.

Nobody knows why Shakespeare decided that Macbeth was an ideal subject for his play; perhaps he just liked the name. What Shakespeare was less keen on was the name of Macbeth's queen, Gruoch, and in the play she is referred to throughout as Lady Macbeth.

It is thought that for his research Shakespeare used the book by 16th-century English historian Ralph Holinshead, *Chronicles of England, Scotland and Ireland*, and that he incorporated many historical inaccuracies of Holinshead's, plus a few of his own into the play. Macbeth was never Thane of Glamis or Thane of Cawdor and Duncan was a relatively young man – not the old king portrayed – and was killed near Elgin fighting the Norse Earl of Orkney and not at Macbeth's castle at Inverness. Macbeth did fight a battle at

Dunsinane in 1054, but was not killed until three years later at another battle at Lumphaman, near Aberdeen, and Banquo does not appear to have been a real historical figure at all. But when you are the world's greatest playwright you are allowed a little poetic licence.

Macbeth was first performed around 1605–6 and has become one of the most popular Shakespearean tragedies. Part of the play's appeal is that it is the shortest of his tragedies, a not insignificant consideration for teachers of students with short attention spans. The numerous films made of *Macbeth* include a 1971 adaptation by Roman Polanski and a Japanese version by Akira Kurosawa in 1957 appropriately entitled *Throne of Blood*, while in the world of opera the play inspired Giuseppe Verdi's *Macbeth* in 1847.

Other than brevity, *Macbeth* has a lot going for it. There is violence, murder, ambition, treachery and guilt, and there are flawed heroes, villains seeking redemption, plenty of action, witchcraft and a wandering ghost. There are also words from the play that have reached far into the national consciousness: 'I bear a charmed life'; 'It is too full o' the milk of human kindness'; 'Double, double toil and trouble'; 'what's done is done'; 'Nothing in his life became him like the leaving it'; 'but only vaulting ambition'; 'full of sound and fury, signifying nothing' and 'Is this a dagger which I see before me?'

– all phrases and expressions in use in everyday life, or if you happen to find a dagger lying around in uncertain light.

Power, ambition and treachery are often associated with the world of politics, and when in 2007 a gloomy, brooding Scot with ambitions to become the main man finally replaced his former ally and leader, and then everything proceeded to go wrong, it was not difficult to draw parallels with Shakespeare's Macbeth. Although sadly for our now former Prime Minister, his torment would last longer than two and a half hours.

HUME IS WHERE THE MIND IS

32 DAVID HUME (1711–76)

In the 18th century Scotland had boasted four established and prestigious universities – St Andrews, Glasgow, Aberdeen and Edinburgh – that had been founded between 1413 and 1583 and having abdicated all interest in politics after the Act of Union of 1707, its people could concentrate on science, engineering and the arts and would change the world in the process. The 18th century is known as the Age of Enlightenment or Age of Reason and Scotland, rather than being on the fringes of change, for once found itself at the centre.

The world was no longer flat, gravity was all the rage and Edinburgh's pubs and taverns were awash with writers and philosophers thinking outside the box, or after a carafe or two of claret completely outside the box.

The first Scottish philosopher to find international fame was the mid-13th-century theologian John Duns Scotus. His writings became respected throughout the Christian world until a couple of centuries later he became discredited and his supporters were given the derogatory nickname of Dunses and Dunsers meaning someone of poor learning. Over the years, Dunses became dunces and applied to people who were slow to learn or not that bright – much to the chagrin of the good people of the Borders' town of Duns, from where John Scotus hailed.

The late-18th century saw the rise in Scotland of the Common Sense School of Philosophy, which attempted to balance critical thought with Christian faith and would have considerable influence throughout Europe and North America. Part of the reason for the rise of the Common Sensers was as a reaction against the work of another 18th-century Scottish philosopher, David Hume, but it would inevitably be the thoughts of Hume that historically and philosophically would prove the most influential, although it was generally agreed that a wee bit of common sense was always desirable.

Hume was born in Edinburgh and when he was only in his twenties published his best-known work *Treatise of Human Nature*, although it was only after Hume's death that it gained its international reputation. Hume had been influenced by the empirical school of thought of English philosopher John Locke that said that knowledge could only be gained through experience.

Hume developed this idea further, stating that the 'science of man is the only solid foundation for the other sciences', and said that what you knew came from experience and observation. In addition, and most controversially for the time, he proposed that Christianity and religion in general was a matter of faith and not a matter of fact; this was radical stuff. Hume never called himself an atheist, he was too much of an agnostic to commit to that extent, but following the great traditions of Scottish jurisprudence everything had to be proven, and God was no exception.

If Hume was not an atheist, there were nevertheless plenty in the Kirk who called him one. Throughout his life his views counted against him by many in the Scottish establishment, and when he applied for positions at the universities of both Edinburgh and Glasgow his CV invariably got lost in the post.

Unbowed, Hume made his name as a historian and wrote his best-selling *History of England*, before settling in France for a while where he was lionised as a great European intellectual and gained the soubriquet, Le Bon David. Eventually Hume returned to Edinburgh where he lived out his life and, albeit considered a dangerous figure, was generally well liked which, on the basis that fifty years earlier anybody holding beliefs such as Hume's might have been executed for heresy, showed how enlightened the Scottish Enlightenment was.

Hume was a bon viveur, enjoying good company and good living as much as he enjoyed the cut and thrust of intellectual debate, be it science, politics or philosophy. He was the Stephen Fry of his century and his monthly quiz nights were one of the highlights of Edinburgh society. As Hume is alleged to have said, 'I have written on all sorts of subjects . . . yet I have no enemies; except, indeed, all the Whigs, all the Tories, and all the Christians.'

David Hume died in 1776 but his reputation as a proponent of sceptical thought and humanitarianism would grow and he is now accepted as one of the greatest European philosophers. Crucial to this was Hume's influence on the 18th-century German philosopher Immanuel Kant and his lesser known and more negative contemporary Immanuel Wont.

EDITOR FOR NEW ENCYCLOPAEDIA REQUIRED, GOOD HYGIENE NOT A PREREQUISITE

33 WILLIAM SMELLIE (1740–95)

With all this Enlightenment going on in 18th-century Scotland, budding scientists, philosophers, educationalists and writers were in desperate need of some good reference books to help to answer the burning questions of the day. With broadband coverage somewhat limited, enterprising Edinburgh printers Colin Macfarquhar and Andrew Bell saw a gap in the market and commissioned an encyclopaedia that would aim to comprehensively cover everything known to man. The man that they hired as editor for this mammoth task was twenty-eight-year-old William Smellie from Edinburgh and between 1768–71 he wrote a hundred weekly instalments that would eventually fill more than 2,500 pages, with accompanying illustrations drawn by Bell. The completed three-volume epic was given the name *Encyclopaedia Britannica*, although it featured a Scottish thistle emblem and, although it was by no means the first encyclopaedia that had been compiled, it quickly gained a reputation as the reference book no quiz night could be without.

Smellie proved a fine writer with a considerable breadth of interests, and what he didn't know he was more than happy to borrow from other texts, although his relative youth may have had some bearing on his entry concerning Woman which consisted of only four words: 'the female of man'.

The encyclopaedia's first edition was the only one that Smellie edited, but the template he created has continued until this day. By 1801 when the supplement to the third edition was published, the tome had expanded to twenty volumes and steadily gained its reputation as the most important reference book in the world. In 1901, after being published in Scotland for 130 years, the *Encyclopaedia Britannica* was sold to an American company. To date there have been fifteen editions and *Britannica* has involved thousands of contributors, including Albert Einstein and Sigmund Freud.

In the 1850s, with *Britannica* already well established, a group of English lexicographers concluded that none of the current English language dictionaries were comprehensive or detailed enough and that a new one should be produced the like of which no bookshelf had ever seen.

Progress was painfully slow, and it was not until 1879 that Oxford University Press agreed to publish such a dictionary under the editorship of James Murray, born in Denholm in the Borders. What was not clear at this stage was how long the project would take, a timescale of ten years was mooted, as every word would have to be thoroughly researched, referenced, checked and verified. Thousands of volunteer researchers gave up their time to work on the dictionary; including one talented, long-term researcher who it turned out was an inmate of an asylum for the criminally insane. But it was Murray, of course, who made all the final decisions about what to include and omit. The first volume, from A to Ant, was published in 1884, with the rest of the 'A's following in 1888 – a very stressful wait for those suffering from 'anxiety'.

Murray worked methodically and meticulously on the dictionary for nearly fifty years, right up until his death in 1915, and even then the work was far from complete. The first edition of the *Oxford English Dictionary* was finally published in 1928 and comprised twelve volumes. Murray was responsible for all of A–D, H–K, O–P and T which meant that he was completely blameless for the infamous omission of 'gullible' from the first edition.

EVERYBODY LOVES RABBIE

34 ROBERT BURNS (1759–96)

The story of Robert Burns is well known. Born in Alloway, near Ayr, in 1756. Inherited his father's farm. Had his first collection of poems published in 1786 in

Kilmarnock to immediate acclaim. Went to Edinburgh, mainly to get away from the aggrieved fathers and husbands of the women he had romanced. Returned to Ayrshire and then to Dumfries where he worked as an excise man. Had numerous children with numerous partners and died in 1796 aged only thirty-seven. Lesser-known facts about Burns include that before the successful publication of the *Kilmarnock Edition* he had been planning to emigrate to Jamaica, as many Scots did, to work as a bookkeeper on a slave plantation, that he was a keen collector and writer of erotic poetry that would be published posthumously as *Merry Muses of Caledonia* and that his nephew, Thomas Burns, was a Presbyterian minister who was one of the founders of the New Zealand city of Dunedin in 1844.

Almost immediately after his death, the life and works of Burns began to be immortalised. The various celebrations of his life would eventually take the form of annual Burns Suppers held on his birthday, 25 January, featuring recitations of some of his most famous poems and songs such as: *A Man's A Man For A' That*; *To A Mouse*; *A Red, Red Rose*; *Ae Fond Kiss, And Then We Sever* and *Tam o' Shanter*. Burns became the national bard of Scotland and Burns Night became the unofficial national day of Scotland, leaving St Andrew's Day far behind in its poetic, drunken, offal-ridden wake. Wherever Scots went they took the poetry and songs of Burns with them and he was celebrated throughout the Scots Diaspora from North America to New Zealand and all points in between.

Although Robert Burns never set foot outside Scotland his name and work became world-famous. His poems and songs of love would find an audience in the 19th century's Romantic movement – and with surprisingly little moral censure of the disproportionately large number of women therein – and his themes of radicalism and humanitarianism would make him an icon of 19th- and 20th-century liberalism. He was alive during the Scottish Enlightenment but he never went to university or ever drank sherry with the luminaries of the day. He was the 'ploughman poet' whom all classes could relate to, then when his work was translated into Russian he became the 'peasant's poet' and enjoyed such popularity under the communist regime that in 1956 he featured on a Soviet stamp – although Stalin clearly misinterpreted one of his most famous lines concerning man's inhumanity to man.

Another of the bard's often-quoted lines, from *To A Mouse*, are '. . . the best-laid schemes o' mice an' men', which John Steinbeck took for the title of his 1937 novel *Of Mice and Men*, a novel which, thanks to the support of countless English teachers, has been read by millions – whether they wanted to or not.

All of this would have been enough to gain Burns a place as one of the world's most famous poets, but what really sets Rabbie apart is *Auld Lang Syne,* the song with which the English-speaking world and a good number of non-English speakers celebrate the advent of the new year. In Japan, the song's melody has been appropriated by a popular song of their own, *Hotaru No Hikari*, which is played at the end of the school day and when shops and restaurants are closing, as a reminder that it is time to return home and enjoy a little Suntory time.

Burns said that he had collected rather than composed *Auld Lang Syne*, but we can assume that he wrote most of the words and in North America the Hogmanay perennial became established as such in the 1930s when New Year concerts featuring Canadian bandleader Guy Lombardo and his orchestra were broadcast on radio throughout America and *Auld Lang Syne* was the first song played after the bells. The song has also featured in a host of Hollywood films, from *It's a Wonderful Life* to *When Harry Met Sally*, even if in the latter Billy Crystal famously, like many million Scots and non-Scots alike, had to confess that he did not actually know what the lyrics meant.

Of course, the perception of Robert Burns as a

humanitarian might never have been very different had he set sail for Jamaica as planned back in 1786. The West Indies brought great wealth to Britain, mainly as a consequence of the Caribbean sugar trade and the concurrent transport and sale of African slaves to work on the cotton and sugar plantations in the New World. By the end of the 18th century there were more than 800 sugar cane plantations and 300,000 slaves in Jamaica. While Bristol and Liverpool became the main hubs of the British slave trade, slave ships did also sail in and out of the Clyde; the trade in cotton and sugar brought huge financial rewards to the Scottish economy and Scots were more than happy to work in both the slave trade and work on the Caribbean plantations as managers, overseers and, as Burns could easily have been, clerks. Far from being innocent bystanders in one of the most shameful periods of British history, by the end of the 18th century as many as 10,000 Scots had gone willingly to work on the Jamaican plantations.

And lest we forget, despite slavery being abolished in the British Isles in 1807 the practice continued elsewhere, thanks in no small part to the influence of Scotland's most powerful 18th-century politician Henry Dundas from Dalkeith. Dundas served as, variously, Home Secretary, First Lord of the Admiralty and Secretary of State for War and the Colonies, and he was more aware than most of the considerable wealth the Empire made as a result of slave labour.

Dundas more than anyone played a crucial role in ensuring that slavery in the colonies would continue unaffected right through to 1833 and it is an inconvenient truth that much of Scotland's wealth during the 18th and early-19th centuries was derived from the slave trade – worth remembering when you next find yourself strolling down the various Jamaica Streets in various Scottish cities.

IF IN DOUBT, MAKE IT UP

35 WALTER SCOTT (1771–1832)

In 1822 Walter Scott was given the task of arranging the festivities to celebrate King George IV's visit to Scotland's capital, which would be the first time a British, English or, for that matter, Scottish monarch had set foot in Scotland for over 170 years. At that time, Scott, born in Edinburgh, was one of the highest profile figures in Scotland. He had made an international reputation as a poet, but poetry alone did not pay the bills and in 1814 he had written his first novel, *Waverley*, which became an international bestseller.

Scott would go on to write a string of novels that not only proved popular throughout Europe but also sparked the public appetite for what would become a lasting genre, the historical novel. Most of his fictional works were set in Scotland but his best-known, *Ivanhoe*, takes place in England in the Middle Ages; Scott's tale of Norman/Saxon rivalry and intrigue and enigmatic characters – Wilfred of Ivanhoe, Richard the Lionheart, Robin Hood – became the template for medieval tales of knights, maidens and derring-do thereafter.

Not everybody was enamoured with Scott's rewriting of history, Mark Twain for one would satirise the whole school of knights in shining armour popular fiction in *A Yankee in the Court of King Arthur* where one of the one main characters was held to utter 'Great Scot!' every time he became overly excited. But if you wanted a fair maiden rescued from an impregnable castle by a chivalrous knight, then Walter Scott was your man.

The poetry, novels and style of writing of Scott placed him in the Romantic movement of early 19th-century Europe and the picture of the romantic Highland hero that he created became ingrained as an image of Scotland the world over. Thus, when he organised George IV's Edinburgh visit the author decided on a spectacle of flags and tartans, with Highland Chiefs and kilts to the fore – and if any clans did not know their tartan or had never had one then entrepreneurial weavers were more than happy to come up with something for the occasion. George IV was persuaded to wear a kilt on the premise that his descent from James VI made him as much a Jacobite king as any previous Stuart. George weighed more than twenty stones (127 kg) and the sight of him wearing pink tights and a kilt which was rather on the short side did lead to unflattering comments – but it was wryly observed that as the Scots had waited so long a for a royal visit, it was nice to see so much of him.

Some fifty years before Scott's *Waverley*, the poet James Macpherson – born near Kingussie in the Highlands – had published, from 1760–65, a series of epic poems known as the *Ossian* cycle. This comprised an extraordinary collection of 3rd-century poetry written by Ossian, son of the legendary Celtic hero Fionn mac Cumhaill, who is given the name Fingal in the poems. Macpherson claimed to have discovered Ossian's original poems and had translated them from Gaelic into English.

The *Ossian* cycle became hugely popular throughout Europe: Napoleon was a great admirer; Johann Wolfgang von Goethe translated a portion of the poems into German and during a visit in 1830 to Staffa and the island's dramatic sea cave – named Fingal's Cave in honour of Ossian's hero – the German composer Felix Mendelssohn was inspired to write his *Hebrides Overture* which, later, became more commonly known as *Fingal's Cave*.

But for all the acclaim the poems attracted there remained considerable doubt about their authenticity. When Samuel Johnson was asked if any man living would be capable of writing such a work he replied, 'Many men, many women, and many children.' Despite Macpherson being unable to produce the 3rd century original sources from which his translations had been made, the authenticity debate went on well into the 19th century before even the most romantic had to accept that Macpherson had been all along the author of what had turned out to be one of the world's greatest literary hoaxes.

While the great German writer Goethe had been so influenced by *Ossian* to go on to write his epic masterpiece *Faust*, Scotland would also make a Faustian pact with its own history. While the Highlands were being cleared of people through mass emigration as first sheep, and then deer replaced them on their land, Lowland Scotland was embracing the romantic, mythical Highland Scotland of Scott and Macpherson as the national identity of the entire country – even though like *Ossian* it was entirely made up. But we probably should not have been too surprised as was it not Walter Scott himself who wrote, 'Oh what a tangled web we weave/When first we practise to deceive!'

GIVE EVERYBODY IN THE FAMILY A BOOK THIS CHRISTMAS

36 DANIEL MACMILLAN (1813–57)

For all Scotland's long, rich history as a nation steeped in literary tradition, and the continuance today of independent publishing houses, there is also a long tradition of Scottish writers, editors and publishers taking their quill southward in pursuit of a wider readership, a higher profile and the chance of a half-decent advance.

Two of the oldest and most influential publishing houses in Britain, Constable and John Murray, both take their names from Scots. Archibald Constable from Fife founded Constable & Co in Edinburgh in 1795, one of the oldest surviving English-language publishers in the world. Constable bought the copyright to *Encyclopaedia Britannica* in 1812 and continued to build on his international reputation as the publisher of Sir Walter Scott's [35] novels, gaining still greater renown in 1897 for Irish writer Bram Stoker's *Dracula*, which brought the world of the undead to millions of readers, much to the horror of the many peaceful, law-abiding vampires who had been quietly going about their business for centuries.

Edinburgh-born John Murray founded John Murray Publishers in London in 1768 and in 1859 John Murray brought out the most important book of the century:

The Origin of Species by Charles Darwin. That year, Murray's was also responsible for the world's first self-help book, helpfully titled *Self Help*, by Samuel Smiles from Haddington, East Lothian, who expressed the virtues of Victorian values and, in particular, the exceedingly presbyterian benefits of thrift, hard work and perseverance. Much of what Smiles advocated was, as far as most Scots were concerned, no more than good, old-fashioned common sense, and why anyone would pay money to read about it was beyond them, but for the rest of the world Smiles's pearls of wisdom were a recipe for success. His book became an international bestseller and inadvertently founded a genre that 150 years later remains one of the most lucrative in the publishing industry. Although as people still continue to buy new self-help books today, this either means that none published in the past 150 years have actually proved especially helpful or, as was once said of the original *Self Help*, they are, 'only suitable for persons suffering from almost complete obliteration of their mental faculties'.

The most famous publishing house to originate in Scotland began life not on the east, but on the west of the country. William Collins was born in Renfrewshire and became a schoolmaster in Glasgow where he founded his business in 1819, producing and selling religious textbooks for schools. The first Collins Dictionary was published in 1824 and in 1841 Collins, who was a firm Presbyterian, gained a licence to print bibles. Ultimately, in 1919, the family moved Collins's headquarters to London and diversified from publishing reference books and atlases into children's books and fiction. Children's fiction included Enid Blyton's *Noddy* series while Agatha Christie, the 20th century's most popular crime writer, was among the stable of writers for adults. Collins published Christie's work from 1926 onward, starting with *The Murder of Roger Ackroyd*, and so beginning a fifty-year era when butlers would be viewed with great suspicion. The Collins family sold the publishing house in 1989, but the name remains incorporated in Harper Collins.

South-west of Glasgow in the Firth of Clyde lies the beautiful island of Arran from where the Macmillan publishing family hail. The family were originally crofters, but the joys of the Brodick Agricultural Show were not enough to keep Daniel Macmillan on the island and as a young man he left to find work in London bookshops. In 1843, Daniel and his younger brother, Alexander, set up their own shops in London and Cambridge and also began publishing textbooks and, later, fiction.

Under the direction of Daniel and Alexander, the publishing house became one of the biggest in the UK with the publication of a range of titles by important authors including Thomas Hardy and Alfred Tennyson and in 1865, *Alice's Adventures Under Ground* [later *in Wonderland*] by Lewis Carroll and, in 1894, *The Jungle Book* by Rudyard Kipling. Macmillan opened its first American office in the US in 1869 and although the UK and US companies separated in 1896 the Macmillan name was retained on both sides of the Atlantic.

Daniel's descendants retained ownership and ran Macmillan in the UK, although Daniel's grandson, Harold, was granted a sabbatical from the family business from 1957–63 so that he could concentrate on his other job – Prime Minister. Having owned the prestigious publishing house for more than 150 years, the Macmillan family finally sold out in 1995, but considering that at the time their most popular authors were Jackie Collins and Wilbur Smith it was perhaps not too surprising that they had no great desire to read another one.

THERE IS A LIGHTHOUSE THAT WILL NEVER GO OUT

37 ROBERT LOUIS STEVENSON (1850–94)

The author, poet and playwright, Robert Louis Stevenson, one of the most popular writers in the history of English literature, was for a time something of a disappointment to his family. His father, Thomas Stevenson, was a lighthouse engineer, his uncle, Alan Stevenson, was a lighthouse engineer, his other uncle, David Stevenson, was a lighthouse engineer and his grandfather and founder of the dynasty, Robert Stevenson, was a lighthouse engineer. They were known as the Lighthouse Stevensons, Britain's and possibly the world's greatest lighthouse-builders. Between them they built more than eighty lighthouses at some of the most inhospitable and treacherous deep-sea locations off the mainland and island coastlines of Scotland thereby preventing the loss at sea of both ships and mariners. Robert Stevenson's masterpiece is the Bell Rock Lighthouse off Arbroath, now automated but still standing nearly 200 years since its construction. Two of Robert Louis Stevenson's cousins also became lighthouse engineers and it was expected that young RLS would share the family's obsession with guillemots, severe gale force nines and the *Shipping Forecast*, but although he was immensely proud of his family's achievements he was destined for a very different career.

One of the reasons why RLS, born in Edinburgh in 1850, was not ideally suited to life in the North Sea was his health, prone to bronchial illness from an early age. He graduated from Edinburgh University with a degree in law – although he barely practised – and spent the rest of his life writing and constantly travelling in as Stevenson said, 'To travel hopefully is a better thing than to arrive'. He journeyed through Britain, Europe and the United States, spent a year travelling around the Pacific, gained a wife and family along the way, before finally settling in the Pacific island of Samoa in 1890. For a while the idyllic lifestyle there seemed to inspire and reinvigorate Stevenson, but in December 1894 he collapsed and died, when he was aged only forty-four. He was buried, as he had asked to be, on top of nearby Vaca Mountain, overlooking the sea, where appropriately he had an excellent view of a lighthouse twinkling in the distance.

Stevenson was a prolific writer of fiction, non-fiction and poetry who in his career wrote two classic Scottish novels, *Kidnapped* and *The Master of Ballantrae*, as well as the perennially popular *A Child's Garden of Verses*. However, the two books that made his international reputation were neither set in Scotland nor written in verse: *The Strange Case of Dr Jekyll and Mr Hyde* and *Treasure Island*.

Published in 1886, *The Strange Case of Dr Jekyll and Mr Hyde* is a novella rather than a novel and became one of the best-selling works of fiction of the time. Stevenson had long been interested in the concept of a split personality and, something of a Bohemian, understood some of the dangers inherent in the repressive morality of the Victorian era; he was aware, too, of the ever-present contradiction in the Scottish character – part dour, hard-working Presbyterian, part romantic, hard-drinking Celt.

The struggle between good and evil within one body has long been a staple theme in literature and on screen. Of the numerous screen adaptations of Stevenson's novella, the most famous is probably the 1931 version for which Frederick March won the Oscar for best actor. And, of course, the expression, 'Jekyll and Hyde' is firmly established as a description for anybody or any situation of a contradictory nature, although for all that Jekyll is associated with 'good' and Hyde with 'bad' we must remember that in Stevenson's original even Dr Jekyll has to accept that for all his alter-ego's less savoury attributes Mr Hyde was the one most likely to buy a round.

While *Jekyll and Hyde* has certainly found fame cinematically, it was not the first live action film to be made by Walt Disney in 1950 and has never been adapted by the Muppets. Both of these achievements belong instead to another Stevenson classic novel and the greatest pirate story ever written, *Treasure Island*. One of the most infamous pirates of all time was Captain William Kidd from Greenock and one of the most popular adventure stories of the 19th century was *The Coral Island* by Edinburgh-born author R. M. Ballantyne so Stevenson was on to a winner when his tale of pirates, treasure, pieces of eight and yo-ho-hoing was published in 1883. And if *Treasure Island* is the world's most famous pirate story, then Long John Silver is the world's most famous one-legged pirate and has inspired generations of young boys and girls to wear eye-patches and walk the gangplank, although thankfully as far as parents were concerned Silver's unhygienic penchant for having a live parrot sitting on his shoulder at all times never really caught on.

THERE'S NO BAKER STREET LIKE HOLMES

38 ARTHUR CONAN DOYLE (1859–1930)

With the advent of cinema and television, popular Scottish writing found a new outlet and a new and wider audience with the work of a host of successful authors being adapted for the screen. John Buchan from Perth, Muriel Spark (born Muriel Camberg) from Edinburgh and Alistair MacLean from Glasgow are three such authors who became household names the world over thanks to the silver screen.

John Buchan had an illustrious career as a British diplomat and Conservative MP that culminated in his appointment as Governor-General of Canada from 1935–40; as governor-general he authorised Canada's declaration of war, in support of Britain, on Germany in 1939. However, Buchan is best remembered for his prodigious literary output and, in particular, his series of five spy thrillers published between 1915–36 and featuring as the main character Scots-born hero, Richard Hannay. Of the five books, by far the best known was the first, *The Thirty-Nine Steps,* which has been made into a feature film three times, although when the first adaptation – Alfred Hitchcock's in 1935 – was so good, you wonder why the others bothered.

Muriel Spark's 1961 novel *The Prime of Miss Jean Brodie* was inspired by her 1930s schooling among the 'crème de la crème' at James Gillespie's School in Edinburgh and the film adaptation was released in 1969. Eponymously titled and shot in Edinburgh, the film starred English actress Maggie Smith, whose mother was born in Glasgow. Smith won many accolades for her performance as the charismatic but delusional Miss Brodie, including the Oscar for best actress and this makes her one of only two actresses to date – the other being Holly Hunter for *The Piano* in 1993 – to have won the award for appearing as a Scot, although with the character played by Hunter being mute, it is generally thought that Smith's Scottish accent was the better of the two.

Scotland's best-selling novelist of the 20th century and, with international sales believed to be more than150 million, probably of all time was Alistair MacLean. After serving in the Royal Navy during World War II, MacLean worked as a schoolteacher in Rutherglen before having his first book published in 1955. His most successful novels have been *The Guns of Navarone* and *Where Eagles Dare* – both were international bestsellers and both were big-budget Hollywood blockbusters despite having intrinsically the same plot: both set in World War II, both featuring a crack, multi-national troop trying to destroy an impregnable Nazi fortress, both having a traitor within their midst and both preferable to going to B & Q on a Bank Holiday Monday.

Notwithstanding all of the above, the award for universal appeal and perennial international popularity for more than a hundred years goes to Arthur Conan Doyle's brilliant detective, Sherlock Holmes. Conan Doyle was born in Edinburgh, where he studied medicine before heading south and setting up in practice in Portsmouth then, in 1891, he moved to London. He began writing short stories at university and in 1912 he published *The Lost World*, a successful novel set in South America where a secret world remains inhabited by prehistoric animals – proving that boys of all ages have been obsessed with dinosaurs for well over a century.

During the early 1900s, Arthur Conan Doyle was a high-profile supporter of spiritualism and campaigned vociferously against perceived miscarriages of justice. His campaigns were marked out by meticulous research and a forensic approach to detail that is, of course, everything you would expect from the creator of the world's most famous private detective.

Sherlock Holmes and Dr Watson first appeared in *A Study in Scarlet* in 1887. In total, over the ensuing 40 years the duo would feature in fifty-six short stories and four detective novels – which became a fiction genre in its own right – in which deduction and science were key.

The long list of actors who have played Sherlock Holmes on the big screen or on television include Jeremy Brett, Michael Caine, Peter Cook, Robert Downey Jr, Charlton Heston, Christopher Lee and Roger Moore, but by far the best-known is the South African star, Basil Rathbone. In the 1939 film *The Hound of the Baskervilles*, Rathbone made the part his own and, alongside Nigel Bruce as Watson and Glasgow-born actress Mary Gordon as Mrs Hudson, he made a further thirteen Holmes's films with the final one released in 1946. After the original Victorian setting of the early films, the action rolled forward to the 1940s with Holmes and Watson having to deal with the Nazis and poor Mrs Hudson having to make the best of wartime rationing.

Sherlock Holmes as created by Doyle was English but the inspiration for the detective was Edinburgh-born professor of medicine and pioneer of forensic science in medicine, Joseph Bell, under whom Doyle had studied at Edinburgh University. Throughout the capital, Bell was renowned for his acute powers of observation and deduction, able to diagnose a patient's ailment and even their occupation before they had said a word; in *A Study In Scarlet* when Homes and Watson meet for the first times this is precisely what Holmes demonstrates to the amazed doctor. When the stories' huge popularity was established and Arthur Conan Doyle said publicly that Bell was the inspiration for his detective, his former lecturer was quietly, and understandably, proud, although he did wish that Doyle had made less references to Holmes' drug-taking as there would always be some students asking Bell where he got his supply.

PETER FLIES, BUT WALT DISNAE

39 J. M. BARRIE (1860–1937)

The Angus town of Kirriemuir, or Kirrie for short, can lay claim to at least two famous Scots. As well as acclaimed author and playwright J. M. Barrie, Kirriemuir was the birthplace of the former hard-drinking lead singer of rock band AC/DC, Bon Scott, and for a long time the town was said to be the birthplace of debonair British film star David Niven. However, despite his full name being James David Graham Niven and serving in the Highland Light Infantry before heading for Hollywood, Niven was actually only as Scottish as Charles Edward Stuart, i.e. not very – and was born in London rather than the Angus town. Coincidentally, Niven played said Charles in one of the least convincing roles in his career in the 1948 Technicolor epic *Bonnie Prince Charlie* which was so bad you ended up rooting for the Redcoats.

James Matthew Barrie wanted to be a writer from an early age and after graduating from Edinburgh University began a career in journalism in London before finding initial success in 1889 with *A Window in Thrums*, a collection of stories based on his hometown. He continued to write stories and also plays, among the best known of which are his comedies, *Quality Street* (1901) and *The Admirable Crichton* (1902).

The character with whom Barrie is most closely associated first appears in a small role in the author's 1901 novel for adults, *The Little White Bird*, before becoming the title character of his 1904 play, *Peter Pan; or The Boy Who Wouldn't Grow Up*; the play, in turn, Barrie adapted as the 1911 novel, *Peter and Wendy*. With the widespread appeal of both the play and the book

Barrie quickly became a wealthy, literary figure, admired by young and old alike the world over.

Barrie's best-loved creation has been adapted many, many times as straight theatre and pantomime, and on the big screen and television. Its most famous adaptation is the 1953 Walt Disney feature-length cartoon which, despite having no decent songs in it, became a lasting, global success – in part because of a very fetching Tinker Bell sprinkling fairy dust wherever she goes. Disney's Tinker Bell became one of the studio's iconic characters, her fame not hindered by the suggestion that she was modelled on Marilyn Monroe. This story was sadly no more than an urban legend, as even the Magic Kingdom was not yet ready to be quite that magical. But Barrie's reputation received a boost, if it were needed, by the sympathetic portrayal of the author in the 2004 film *Finding Neverland* with Johnny Depp gaining an Oscar nomination for playing the writer, and with a Scottish accent so good that Ewan McGregor vowed he would never again play a Scottish character.

While Peter Pan had first appeared in *The Little White Bird*, it was not until the stage play that Captain James Hook had his chance to shine and become one of fiction's most infamous pirates. The eldest of the Darling children is Wendy, a name that is said to have been invented by Barrie after hearing the young daughter of a friend refer to him as, 'fwendy', or my 'friendy-wendy'. The Darling children fly away with Peter to Neverland or Never Never Land, but there is absolutely no truth in the story that their absence was not noticed by their father as he was too busy dealing with the economy at the time.

J. M. Barrie became a baronet in 1913 and he was awarded the Order of Merit in 1922. He was Lord Rector of St Andrew's University and Chancellor in 1931 of Edinburgh University. His marriage was unhappy, ending in divorce, and he never had children of his own, and gave all the royalties from *Peter Pan* in perpetuity to the Great Ormond Street Hospital for Children in London.

The year before J. M. Barrie was born in Kirriemuir, Kenneth Grahame was born in Edinburgh. Grahame moved to England at an early age where he grew up outside London on the banks of the River Thames. He spent thirty years working for the Bank of England and in his spare time wrote children's stories, including *The Wind in the Willows*, published in 1908 and which has sold more than twenty-five million copies worldwide, making it one the best-selling children's books of all time.

The book tells of Toad, Ratty (who is, confusingly, actually a water vole), Badger and Mole and the friendship that binds the four together. Mr Toad has become one of the best-loved characters in children's literature, despite being arrogant, conceited, selfish, obscenely wealthy and believing himself to be completely above the law, and one can only wonder where Kenneth Grahame, with his thirty years of banking experience, gained his inspiration.

MEN ARE FROM MARKINCH, APPARENTLY WOMEN ARE, TOO

40 MARIE CARMICHAEL STOPES (1880–1957)

I am afraid there is not a large number of women in my list of 100 Scots who changed the world. In fact, the number is so embarrassingly low that I am not even going to tell you what it is. You will have to read the rest of the book before, undoubtedly, becoming downright black-affronted at the inadequate level of women represented and the complete absence of anyone called Lulu. It is a sad, but true, fact that for a country which since the Reformation has put so much importance on education, that in the 18th century was at the very epicentre of the Enlightenment and in the early 20th century was the midwife at the birth of the gender-equal Labour Party, that while Scottish men have been changing the world, Scottish women have been stuck at home bringing up the bairns and making the tea.

Two women who don't who make the list, because they were not born in Scotland, are Kate Shepherd (born Catherine Malcolm) and Elsie Maud Inglis. Both were born of Scottish parents, but in New Zealand and India respectively: Shepherd led the campaign for women's suffrage in New Zealand – the first country in the world to concede universal suffrage, in 1893, while Inglis, famous doctor and suffragette, having moved to Edinburgh when she was ten, established a maternity hospital staffed by women for the destitute who could not receive treatment anywhere else. During World War I she set up women-only field hospitals to care for Allied troops on the front line and her fourteen units of the Scottish Women's Hospital Unit saved countless numbers of lives on the battlefields of France, Serbia and Russia.

Marie Stopes, however, *was* born in Scotland. She was born in Edinburgh and her mother was the Scottish writer and advocate for women's rights, Charlotte Carmichael Stopes, and her father, Hugh, was a scientist. Reflecting the influence of both parents, Marie Stopes graduated with distinction from University College, London in 1905 where she studied botany and she, also, joined the suffragette movement. Stopes married in 1911, but the union was not successful and was annulled in 1914.

Determined that other newly married women, and for that matter men, should not enter matrimony so ill-informed as she had been, she decided to write a useful and informative guide on the subject which was published in 1918. The book was called *Married Love*, and in an era when sex education was virtually non-existent was revolutionary in its frank, no-nonsense descriptions of the sexual organs, intercourse, ovulation and the menstrual cycle – and, most shockingly at the time, that women had their own sex-drive; rather than lying back and thinking of England (or Scotland for that matter) they were thinking of something far more enjoyable.

Stopes expressed the sage advice that quality was always better than quantity and that the best sex was experienced when both participants were mutually fulfilled. Furthermore, in her opinion, the best sex in marriage was usually experienced when after a period of abstinence of around ten days there followed three days of, as she put it, 'regular unions'. So controversial were her points of view that Stopes had been turned down by numerous publishers before the book finally saw the light of the day. It was banned in the US where it was considered obscene, but through word of mouth soon found a ready audience of British women and by 1930 had sold more than 750,000 copies, with 750,000 British husbands probably none the wiser as to why every second weekend they were not allowed out of the bedroom.

Emboldened by her success and impervious to criticism from the Church, politicians and the unenlightened public, Stopes moved on to the matter of birth control and the right of women to avoid unwanted pregnancy. In London, in 1921, she established the first family planning clinic in Britain where an all-female staff of doctors and nurses distributed advice, information and contraception to any married woman who wanted it. When she stated publicly that birth control should be targeted at the poor in order to stop them breeding, many in the middle classes agreed with her but were, of course, too polite to say so. The need for Stopes's advice is summed up in a chant from the time:

> Jeanie, Jeanie full of hopes
> Read a book by Marie Stopes
> But to judge from her condition
> Must have read the wrong edition.

Marie Stopes died in 1958, but her name lives on with more than 500 Marie Stopes International Clinics in forty-three countries – with the majority in Africa and Asia – providing family planning advice, contraception for women, vasectomy and safe abortion for millions of clients. Through her championing of women's rights across the board Stopes became and remains an international icon, even in her conservative hometown of Edinburgh, where women are of course no longer expected to make the tea, as it is generally assumed that you will have already partaken.

International Scots

When in April 2010 in retaliation to what they saw as unfair financial repayments being imposed on them the Icelanders unleashed the Eyjafjallajoekull volcano, a huge cloud of ash grounded all aircraft throughout the UK and large parts of Europe. This of course caused widespread international disruption, but it could be argued that Scotland suffered more than most, not just because its geographical proximity to Iceland meant Scottish airports were among those first affected, but because after centuries of Scots being able to emigrate at the drop of a hat and make a new life all over the world, the psychological trauma of not being able to leave the country, was almost too much to bear – even if it was only for 72 hours

FREE BY 1783

41 JAMES WILSON (1742–98)

1776 was an important year in the Scottish Enlightenment – Adam Smith [23] published the *Wealth of Nations* and David Hume [32] died without recanting any of his sceptical views about Christianity. It was said around that time that you could wander up the Royal Mile in Edinburgh and bump into at least a dozen geniuses. But at the same time as Scots were changing the worlds of science and philosophy at home, Scots were also asserting their influence on important and revolutionary world events abroad.

According to a recent US census six million Americans, or two percent of the population, claim Scottish origins. But take into account the Scots who have emigrated over the centuries and the true figure – although almost impossible to verify – is believed to be closer to twenty million, and this does not include the estimated twenty million Americans who have Scotch-Irish ancestry.

When the first US census was taken in 1790 the overall population was just under four million, of which 300,000 were Scotch-Irish while only 150,000 were of pure Scottish descent, but those 150,000 had already had a disproportionate influence on the history of the fledgling nation.

Aberdeenshire-born Rev James Blair founded one of America's oldest and most prestigious educational establishments, the College of William and Mary in Williamsburg, Virginia, in 1693. US President Thomas Jefferson was educated there and was considerably influenced by his professor, William Small, from Angus. When on the 4 July 1776 at Philadelphia the Second Continental Congress declared the independence of the Thirteen Colonies from Great Britain, it was Jefferson, along with another future US President, James Madison, who were the principal authors of the Declaration of Independence to which two of the original signatories were Scots: John Witherspoon, born in Gifford, East Lothian, and James Wilson, born near Cupar in Fife.

Witherspoon was a Church of Scotland minister who had become the distinguished and respected President of the Presbyterian College of Princeton in New Jersey, the forerunner to the prestigious Princeton University, where one of his students was the afore-mentioned James Madison.

Wilson emigrated to Pennsylvania at the age of twenty-four where he established himself as a prominent lawyer. He was one of the first supporters of independence and published a pamphlet to that effect in 1774. He signed the Declaration in 1776 as a representative of Pennsylvania and, with the war safely won, became a delegate at the Constitutional Convention in Philadelphia in 1787 and, two years later, was one of the first Justices elected to the Supreme Court. Wilson played a prominent part in the Constitutional Convention that eventually decided to accept the bulk of James Madison's proposal to have a presidency with a separate, two-house congress as the legislature and a supreme court as the judiciary. Wilson was elected to the Committee of Detail which came up with the final draft and it was Wilson who took on the job of writing it – and, because he was making the revisions, he made sure that all the parts he agreed with stayed in, including the stipulation that the president and the senate should be elected rather than appointed. Wilson's draft was adopted by the Convention and a final version was ratified by individual States in 1788 and that became the Constitution of the United States.

James Wilson lived and worked in Philadelphia, a city at the very forefront of the American Revolution and capital of the independent United States from 1790–1800. However, in America, as it is in the land of his birth, the name of James Wilson is mostly forgotten. It was not Wilson who wrote the introduction to the Constitution which begins, 'We, the people . . .', but those words are highly appropriate to the democratic principles he advocated. Although, there was some disappointment when it emerged that 'We, the people' was not, in fact, a new, improved game console.

FURRY BOOTS COUNTRY

42 JAMES MCGILL (1744–1813)

In the last Canadian census of 2001 just under five million people – fifteen percent – claimed Scottish ancestry: two million lived in Ontario, 800,000 lived in British Columbia, 600,000 lived in Alberta and the MacRury family from Barra were still panning for gold in the Yukon Territory. Whether it be the wide, open spaces, the bitterly cold winters or always being in the shadow of a wealthier, more powerful neighbour and generally ignored by the rest of the world – the Scots have always felt quite at home in Canada.

The journeys of Alexander Mackenzie [44] to the north and west coasts of what would become Canada are but one of the many examples of the Scots' mapping the geography of the second largest country in the world. During the 18th century and the first half of the 19th, Europe's interest in what were then America's northern and western territories was the burgeoning fur trade and Scots were at the forefront in the pursuit of beaver. Scottish traders, agents, managers and trappers came from both the Lowlands and Highlands, but particular reference must be made to the large number who came from the Hebrides and from Orkney who, with regard to the weather conditions at home and their low centre of gravity were ideally suited to spend months battling the Arctic winds.

The Hudson's Bay Company was the major trader and one of its employees was Orkney-born doctor and explorer John Rae. Rae was the greatest Arctic explorer of his day and by 1854, having completed a mission to find out what had happened to John Franklin's ill-fated British expedition in search of the Northwest Passage, was the man who finally mapped the last sections of the Passage, although it was not recognised as such at the time. With fewer than ten men and travelling overland, Rae had achieved what had until then eluded larger and ostensibly better-prepared expeditions for centuries. He was, however, dishonoured at home by Franklin's supporters, who didn't care to accept the truths Rae was obliged to report about the fate that had befallen Franklin and his men with many in the establishment vowing never to eat Orkney fudge again.

In 1760 the British defeated and captured the French territories in North America and three years later East Lothian-born James Murray was appointed the first governor of British-run Quebec. Murray – whose nephew was Patrick Ferguson [53], inventor of the breech-loading rifle – was determined to gain the support of the French-speaking Quebecois and, much to the annoyance of the English-speaking population, protected the status of their language, religion, legal system and business interests.

The remote north and the west regions were the province of the fur traders who ran operations pretty much as they pleased. In 1821 the two main, rival traders, the Hudson's Bay Company and the North West Company, merged and the man appointed North American manager of the new company was Dingwall-born George Simpson. Simpson thus became, effectively, governor of an area covering three million square miles that included most of modern day Manitoba, Saskatchewan and British Columbia. Such an onerous responsibility might have proved too much for some, but Simpson took it in his stride – if you could cope with a Saturday night in Dingwall, then you could cope with most things.

For many of the fur traders, Quebec's largest city and port, Montreal, became their base and it was there that Glasgow-born fur trader and merchant James McGill made his home. McGill became one of the wealthiest men in Quebec and decided that in his will he would give some of his money back to the people of North

America. He left a bequest of money and land to build a university in Montreal and in 1821, eight years after his death, McGill College, later McGill University, was founded. The university became the most prestigious in Canada and is at time of writing ranked in the top twenty in the world. The motto of McGill University is 'by hard work, all things increase and grow', the very same motto of lonely Scottish trappers in the long winters of the Arctic tundra.

NEIGHBOURS, EVERYBODY LOVES REHABILITATED NEIGHBOURS

43 LACHLAN MACQUARIE (1762–1824)

Ulva is one Scotland's least known and least visited inhabited islands. Privately owned, it lies to the west of Mull, is only seven miles (11 km) long by two miles (3 km) wide and supports some twenty inhabitants. And unlike islands such as Skye and Iona, which have seen an upsurge in popularity in recent years as first names, there has, perhaps understandably, not been any trend for Scottish parents calling their baby girls Ulva. Yet it was on this small, Inner Hebridean island that the man who would become the 'Father of Australia' was born.

Lachlan Macquarie joined the army at the age of fourteen and worked his way up the ranks, serving in different posts around the world until 1810 when he was appointed as the fifth Governor-General of New South Wales. It had been twenty-two years since the First Fleet had sailed into Sydney Cove and founded the first British settlement in Australia. The British still smarting from the loss of the Thirteen Colonies in America were keen to establish somewhere new – even it was 8,000 miles (13,000 km) away. They had come up with a cunning plan to use mostly convict labour to build the colony and across Britain and Ireland any stonemasons, craftsmen and labourers who had the misfortune to be caught stealing a bread roll found themselves sentenced to seven years labour on the other side of the world – and heaven help you if you pilfered an entire loaf. As Scottish law was less repressive than English law, transportation was reserved for only the worst Scots miscreants, therefore making them the people you would least have liked to meet on a Saturday night in downtown Sydney, a situation that has arguably continued to this day.

The early years were tough for the European settlers in Sydney and Tasmania, and Macquarie's predecessor had faced a military revolt in 1808 – although considering his predecessor was a certain William Bligh this should only have been expected; Macquarie had been appointed to restore order in the colony. There was no legislative government in New South Wales and with his superiors far away in London, Macquarie had absolute power to run the colony as he saw fit.

Macquarie was determined to transform the reputation of New South Wales as the biggest open prison in the world into that of a proper, well-run colony and, with this in mind, he gave the convicts who had completed their seven-year sentences the same rights as the free settlers and backed this up by appointing ex-cons to positions of authority.

He followed an expansionist policy that included the crossing of the Blue Mountains to found the first inland settlement of Bathurst, as well as explorations north into what would become Queensland. He also oversaw the transformation of Sydney from a village into a city with Macquarie Street becoming its most exclusive address; the street would, later, be the location for the New South Wales Parliament and, in 1817, for the Bank of New South Wales, the first bank in Australia.

Macquarie's reforms and autocratic rule made him enemies. He was forced to resign in 1817 and returned to Britain in 1822. He did, however, have the satisfaction of leaving a thriving, growing colony of 40,000 settlers. Having said that, all this expansion came at the expense of the native population whose land and water were taken from them and, in return, all they received from the British was smallpox. Thousands of Aboriginals who had the misfortune to come into contact with the settlers died during his governorship and, although Macquarie was at times personally sympathetic to the plight of the Aborigines, his responsibility was first and foremost to the settlers and the decimation of the indigenous Australians continued in ever-greater numbers.

Captain James Cook, the son of a Scottish farm labourer but born in Yorkshire, coined the name New South Wales when he sailed into Botany Bay in 1770 but some forty years later *Terra Australis*, Latin for 'southern land', was proposed as a more fitting – and less Welsh – alternative for the continent and in 1817 Macquarie became the first governor-general to officially adopt the name Australia. Lachlan Macquarie, the Father of Australia, established the foundations of the nation that we know today as opposed to a convict colony; Macquarie University and Macquarie Bank are among the numerous buildings and places that bear his name. He died in 1824 and a mausoleum was raised to his memory on the island of Mull – just a few miles over the water from Ulva – and many grateful Australians have made the long pilgrimage to Mull to pay their respects and steal the flowers.

DEEP-FRY ME A RIVER

44 ALEXANDER MACKENZIE (C1755–1820)

In 1621 James VI [11] granted a charter to William Alexander, the Earl of Stirling, for the founding of a Scottish colony on a peninsula on the south-east coast of what is now Canada. The peninsula was given the name Nova Scotia, Latin for new Scotland, and the first Scots settlement began in 1629. The climate of Nova Scotia is often compared to the climate of Scotland and the new settlers felt right at home when they encountered the region's speciality: freezing hail. This Caledonian enclave did not last long however, and in 1631 Nova Scotia was handed over to the French and would not be returned to the by then British, rather than English or Scottish, until 1713. The name Nova Scotia was retained, although it was not until the late-18th century that large numbers of Scots began to settle in Cape Breton Island, just to the north-east of the Nova Scotia mainland.

More than 50,000 Highlanders immigrated to Cape Breton, taking their Gaelic language and traditional music with them, and it was around this time that the island's native moose population disappeared, although it remains unclear whether there was any connection between this and the incessant fiddle playing.

Nova Scotia became a self-governing British colony in 1848 and joined the new Canadian Confederation in 1867. Today the population of Nova Scotia is one million, and one third of these claim Scottish descent. The region still boasts 1,000 Gaelic speakers and is the only Gaelic-speaking community outside Scotland, although as in Scotland numbers are in steady decline. In recent years tourism has become an important component of the Nova Scotia economy, with the region's Scottish heritage proving to be one of the major draws, and visitors will spend days in custom-built hides in the hope of hearing Gaelic speakers talk to each other.

By the end of the 18th century the European settlement in the north and west of British North America was concentrated around the fur trade. The largest trader was the Hudson's Bay Company whose headquarters was in London but in North America was managed mostly by Scots. In 1779 the rival North West Company was co-founded by Inverness-born Simon McTavish and eight years later a Gaelic-speaking trader, Alexander Mackenzie, joined the company. Mackenzie was probably born in Stornoway on the Isle of Lewis, although some have said Inverness, and emigrated first to New York and then Montreal, which was where he became involved in the fur business.

He was given the job of exploring the country's

western territories with a view to the company trading there and in 1789 he set off with a party of 8 native men, 4 of their wives and 2 birch-bark canoes. The party hoped to find the legendary Northwest Passage but, instead, Mackenzie found the second longest river in North America and the longest of what is now Canada before reaching the Arctic Ocean. As the whole point of the journey had been to find a viable passage connecting the Arctic to the Pacific, Mackenzie was understandably gutted at having come all that way to end up at the wrong ocean. He named the river Disappointment – probably one of the milder of the words he used at the time – and only much later was Disappointment River renamed Mackenzie River in his honour.

Undeterred in 1792 he set off again with another party and, this time, headed west but via a more southerly route. He crossed the Rocky Mountains and eventually arrived on the Pacific coast of what is now British Columbia – the first European to cross the continent of North America north of Mexico from coast to coast.

The exploits of Mackenzie did not gain an immediate benefit to the North American fur trade, but would inspire other explorers, traders and politicians throughout the continent to look way out west. Mackenzie could not have undertaken any of his major explorations without the direct help and guidance he received from the native population, although had they known that it would lead to millions of other Europeans turning up in their country, they would have left him to the bears.

WALTZING ST KILDA

45 THOMAS BRISBANE (1773–1860)

Lachlan Macquarie [43] is known as the Father of Australia but he was not the first Scot to be Governor-General of New South Wales. John Hunter from Leith had sailed with the First Fleet in 1788 and was appointed the second governor-general from 1795–1800; the Hunter River, north of Sydney, is named after him and the region around it is has become world-famous as the Hunter Valley. And the longest river in Australia is the mighty Murray, more than 1,572 (2,530 km) miles long, named after Perth-born George Murray who was Secretary of State for War and the Colonies from 1828 to 1820 when Europeans first discovered it.

By the 1820s the British had been in Australia for more than forty years but had only travelled and settled in a fraction of the continent. Western Australia had been almost completely ignored; it was more than 2,000 miles (3,219 km) from Sydney to the western coast and who on earth would want to live in such a remote place? Step forward Scottish admiral and administrator James Stirling from near Coatbridge in Lanarkshire who along with Perthshire-born botanist Charles Fraser explored the western region in 1827 and found a location they thought would be ideal for settlement.

Two years later, Stirling returned as the first governor of the Swan River Colony, and by extension the first governor of Western Australia, which would, later, become the cities of Freemantle and Perth. Stirling named the latter not after himself but for the aforementioned Secretary of State George Murray's Perthshire roots: Murray, it must be said, has a disproportionately major influence on Australian place-names considering he never went further east than Denmark. Stirling may have decided not to call the new settlement after himself because, contrary to the glowing reports that he and Fraser had written about it, Perth was not especially fertile, much to the disappointment of the initial settlers who grumbled that as far as Scottish place names were concerned Coatbridge, or even Airdrie, might have been more appropriate. It was not until the 1890s with the discovery of gold in western Australia that Perth began to grow any larger than a small town. The population was only 5,000 in 1881, but had grown to 250,000 by 1947 through the prosperity brought by industry, mining and agriculture. Today, Perth – population some 1.6 million – is the fourth largest city in Australia; forty times the size of its Scottish counterpart and recognised as one of the best cities in the world to live – although having said all that, it is still bloody miles from anywhere.

The third largest city is Brisbane – capital of Queensland, population 1.7 million – named after Thomas Brisbane, born near Largs in Ayrshire and who had a distinguished career in the British Army before succeeding Macquarie in 1821 as Governor-General of New South Wales. Brisbane was in office for some four years until, like his predecessor, he was recalled on account of his liberal policies and returned to Scotland, where he passed the rest of his years on his country estate.

During his short-lived governorship, Brisbane sent an expedition north into what is now Queensland in search of somewhere on the coast to start up a new penal settlement. Established in 1824, the new settlement was named Brisbane, as was the river upon which it was built – and Brisbane visited to find convicts and guards alike equally disgruntled at the lack of surfing facilities. Brisbane ceased being a convict settlement in 1842 and seven years later became the capital of Queensland. A major trading port during the 20th century, the city grew rapidly and continued to do so as the tourists' gateway to the Gold Coast to the south, to Cairns and the Great Barrier Reef to the north and as the default destination for all departing *Neighbours* characters.

There are numerous other Scottish-influenced place-names in Australia, including the county's sugar cane capital, Mackay in Queensland named after Inverness-born explorer John Mackay and the 1,700-mile (2,735 km) Stuart Highway, which runs from Darwin in the Northern Territory to the South Australia coast and was named after Fife-born explorer John McDouall Stuart. It was Stuart who in 1862 became the first European to cross Australia from south to north, although the expeditions took a terrible toll on his health and he died just four years later aged just over fifty, having sadly concluded that it had been an awful lot of effort just to see a big red rock.

The settlers were, of course, taking land from the native Australians and simply imposing new names on regions, rivers and so on as they pleased. Recent years have seen the first, albeit belated, attempts by the Australian government to make some recompense for the cruel injustices inflicted on the indigenous people ever since the first Europeans arrived, but there is a long way to go.

It is also worth noting when considering the naming question that some of those chosen by second- and third-generation European Australians have been painstakingly literal. A prime example was following the discovery in Western Australia of a great sandy desert the size of Japan which, after much deliberation, was given the name, The Great Sandy Desert or, to give it its full name, Strewth Mate, That's A Great Sandy Desert.

I WENT TO NEW ZEALAND FOR MY HOLIDAYS, BUT IT WAS CLOSED

46 JAMES BUSBY (1802–71)

The popularity of New Zealand for emigrating Scots has always been a conundrum for those who study demographics. Was the main attraction the fact that New Zealand is 12,000 miles (19,312 km) from Scotland and therefore, in the days before air travel, once you'd got there you were very unlikely to return? Or was it that much of New Zealand resembles a larger, quieter version of Perthshire, if such a thing is possible, which makes the Scots feel right at home?

Whatever the reason, the Scots settled throughout North Island and South Island. When the British authorities founded Auckland, in 1840, the first substantial group of settlers were 500 Scots from Paisley, who arrived two years later only to be faced with little in the way of employment – so no change there. Many Scots' settlers discovered a special affinity for two of the coldest, wettest districts, Otago and Southland in South Island. In Otago, the university city of Dunedin – an anglicised version of the Gaelic form of Edinburgh – was founded in 1844 by a contingent of Wee Frees. Robert Burns's nephew, the Rev Thomas Burns, was one of them and, while he could not compete as far as poetry, drinking and womanising were concerned, he was more than a match for his uncle when it came to shooting possums.

As with many things in British imperial history it was Captain James Cook who was at the beginning of it all. On the same expedition on the *Endeavour*, a few months before making landfall in Australia Cook laid anchor off the North Island of what he would call New Zealand in 1769, rowed ashore and claimed it for Britain. British whalers, traders and missionaries began to settle in New Zealand from the early nineteenth century onwards, with opposition from the native Maori population reduced by the strategy of giving them muskets to fight each other rather than the British interlopers.

With little, if any, control and the French showing an interest, the British Government decided to appoint an Official British Resident in 1833 to help establish some order: Edinburgh-born James Busby, who had emigrated with his family originally to Australia. While Busby was not in a position to exert much control over the Europeans in New Zealand, he was more successful in persuading the Maori tribes to work together and make the islands more peaceful. However his diplomatic skills in uniting the Maoris resulted in making the country even more appealing to his superiors back in Britain who now decided to officially colonise it and thus, in 1840, Busby found himself co-author of a treaty between the British Government and a number of Maori chiefs from the North Island. The Treaty of Waitangi was signed on 6 February 1840 in a marquee on the front lawn of Busby's house – at Waitangi – in the Bay of Islands and proved one of the most important and controversial events in New Zealand's history.

There were two versions of the treaty, one in English and one in Maori. The British took the treaty as an official agreement that New Zealand would become a British colony governed by the Crown and, in return, Britain would recognise Maori land-rights and the right of Maoris to become British subjects. The Maoris, on the other hand, understood that they were agreeing only to a British presence in New Zealand and unbeknown to them, and with Busby no longer Resident, in 1841 New Zealand became a British colony.

The first Parliament met in 1854 and three Scottish-born men held office as Prime Minister: Edward Stafford from Edinburgh served three terms between 1856 and '72; Robert Stout from Shetland served two terms between 1884 and '87 and Peter Fraser, born near Tain in the Highlands, was Prime Minister from 1940–49 and led his country throughout World War II when more than 100,000 New Zealanders served, mostly in Europe.

When he left his position as Resident in 1840, Busby stayed on in New Zealand and, being careful to keep clear of any disgruntled Maoris and having thought better of his promise to get a full body tattoo, he resumed his original vocation as a winemaker. Before immigrating to Australia he had studied viticulture and winemaking in France, and he was not about to allow being several thousand miles from the nearest vineyard stop him indulging in a cheeky little merlot. In Australia in 1824 he had planted the first vineyard in Hunter Valley and a year later published the country's first book on wine. Busby was not the first to grow vines in either Australia or New Zealand, but it was his decision to grow vines in what would become one of Australia's greatest winegrowing regions, and his unstinting promotion of the merits of the Aussie wine has earned him the moniker, the Father of the Australian, and for that matter, the New Zealand Wine Industry.

It would not be until the 1980s that Australian and New Zealand wine would see the beginning of a dramatic increase in the consumption and popularity internationally and today Australia is the fourth largest wine exporter in the world. James Busby would no doubt be delighted to know that not only is Britain Australia's largest export market, but that we buy more Aussie wine than we do French or Italian, presumably on the basis of the more chardonnay you drink, the less you seem to care what it actually tastes like.

NO SNOW AGAIN IN AFRICA THIS CHRISTMAS TIME, I PRESUME

47 DAVID LIVINGSTONE (1813–73)

In the late-18th century Mungo Park, born near Selkirk, and James Bruce, from Stirling, ventured deep into the African interior where they claimed credit for the discovery of the sources of the Niger and the Blue Nile respectively, much to the bemusement of the local people who took them there.

It is with great surprise that one learns that the town of Blantyre in South Lanarkshire has a population of around 17,000 while its namesake in the African republic of Malawi – which gained independence from Britain in 1964 – has a population of more than 750,000. Are there really 17,000 people living in the Scottish Blantyre?

What do they all do there? The African Blantyre on the other hand is Malawi's largest city and its industrial and financial centre. It was founded by Church of Scotland missionaries in 1876 and takes its name from David Livingstone, who was born in the Scottish Blantyre in 1813. Livingstone had actually left the Church of Scotland as a young man to join the rival, and smaller, Congregational Church, but the Presbyterians in Africa did not hold that against him.

In Africa Livingstone would meet fellow Scots missionary, Robert Moffat from East Lothian. Moffat lived in what is now the north of South Africa for fifty years and it was he who suggested to Livingstone that he travel farther north into central Africa, which was then – as far as Europeans were concerned – unknown. If this was a cunning plan to split up the budding relationship between Livingstone and Moffat's daughter, Mary, it didn't work as Mary married Livingstone and decided to go with him instead.

From 1845–73, with only two brief returns home for good behaviour, Livingstone spent the rest of his life in deepest, darkest Africa, endlessly travelling from lake to lake and river to river, and much to the consternation of the ever decreasing members of his party who, in the words of his Scottish doctor, John Kirk from near Arbroath, could only conclude that, 'Dr Livingstone is out of his mind'. In 1869 Livingstone had been out of circulation for four years and the explorer/journalist Henry Stanley was sent to try and find him by the *New York Herald*. In November 1871 the two men finally met on the shores of Lake Tanganyika – in what is now Tanzania – whereupon Stanley uttered the immortal words, 'Dr Livingstone, I presume?', and Livingstone responded by informing Stanley that all appointments had to be made between nine and ten in the morning.

As far as his official job as a medical missionary was concerned Livingstone was less than successful as in all the years he spent in Africa he does not appear to have either converted or healed anybody. But as an explorer he became a Victorian hero of worldwide renown. His exploits would inspire the powers in Europe to look to Africa as a continent rich in opportunities, and the 1880s saw the beginning of the Scramble for Africa with Britain, France, Germany, Italy, Belgium and Portugal dividing up the continent between them; with the local consultation process restricted solely to what European language the native population would prefer.

William Mackinnon from Campbeltown in Argyll founded the British East Africa Company in 1888. He had made his fortune in British India as a shipping magnate before looking for new opportunities on the other side of the Indian Ocean, and he found them when his company between 1888 and 1890 leased the stretch of African coastline of what would become Kenya and Uganda, that would become a British Protectorate in 1895, so beginning seventy years of colonial rule.

Mackinnon was not the first Scot to have colonial aspiration in Africa. In 1806 General David Baird, from Haddington in East Lothian, successfully captured Cape Town from the Dutch so beginning the long history of the Cape Colony as a British colony, and over the next century thousands of Scots would travel to South Africa to work as administrators, doctors, farmers, engineers, miners and big game-hunters – after shooting grouse in the Highlands, missing an elephant was nigh-on impossible.

One such Scot was Leander Starr Jameson from Edinburgh, whose life was to prove as colourful as his name. A doctor by profession, Jameson immigrated to Africa where he became involved in colonial-style politics during the increasing tension between the British and the Dutch Boers over control of the lucrative diamond and gold mines.

In 1895 Jameson led 500 British in an attack – the Jameson Raid – on the Boer Republic of Transvaal. The raid had been tacitly supported, but not officially sanctioned by the British authorities, and would prove completely ineffective with Jameson being put in prison for fifteen months. But the maverick doctor still had friends in high places and before long his failure would be turned into a glorious example of the brave, gallant British fighting the beastly Boers and Jameson returned to politics as prime minister of the Cape Colony from 1904–08 – an inspirational Starr for those imperial times of stiff upper lips, unbending moustaches, a sun that never sets and a Britishness that remains resolute in the face of adversity – as long as the locals knew their place and remembered to keep the tonic water chilled at just the right temperature.

CAREFUL NOW . . .

48 JOHN A. MACDONALD (1815–91)

In 2002 the BBC asked the public to vote for their Greatest Briton of all time and the findings did not make comfortable reading for the Scots. The highest-ranking Scot was the man who discovered penicillin, Alexander Fleming [67], who sneaked in at number twenty, well below Elizabeth I (who chopped a Scottish queen's head off), Oliver Cromwell (who conquered and governed Scotland in the 1650s) and Margaret Thatcher (who was so unpopular north of the border that every single Scottish Conservative MP was voted out of office). However, if the Scots' ranking in the BBC poll was disappointing, there was better news when Canada replicated the format and three out of the top ten Greatest Canadians hailed from Scotland. This is perhaps not so surprising when you look at the numbers: five million Canadians, or fifteen percent of the total population – equivalent of the total current population of Scotland – claim Scottish origin.

The three Scots featured in Canada's top ten were Tommy Douglas, Alexander Graham Bell and John A. MacDonald. Douglas, born in Falkirk, immigrated with his family to Manitoba and was a Baptist minister before entering politics. In 1944 he became premier of Saskatchewan on a social democratic platform and remained in power until 1961. His main achievement was the introduction of a universal health insurance scheme in Saskatchewan in 1962 that was adopted countrywide in 1966 and earned him the title, Father of Medicare. Edinburgh-born Alexander Graham Bell [96] emigrated to Canada at the age of twenty-three and was, of course, the inventor of the telephone. And the third Scot was John A. MacDonald, with the A standing for Alexander, Canada's first ever Prime Minister.

MacDonald was born in Glasgow and, aged five, immigrated with his family to Canada where he qualified as a lawyer before entering politics. He became Canada's very first prime minister in 1867–73 and again from 1878 until his death – just three months after being re-elected to office – in 1891. He gained a reputation as a schemer, a drinker and someone for whom the whiff of scandal and corruption were never far away – an archetypal West of Scotland politician some might say – nevertheless, his premiership laid the political foundations from which the dominion grew toward the Canada of today.

The loss of the Thirteen Colonies to the United States in 1783 at the end of the American War of Independence was a bitter blow for Britain, but as they say, if you love someone sometimes you have to let them go. With no wish to repeat such a loss elsewhere in North America, the British came up with the fiendishly clever strategy of uniting French-speaking Quebec with English-speaking Ontario in a new Province of Canada where nobody would have a majority and nothing would ever get done. But if anybody knows how to make the most out of devolution, it is the Scots.

The two leaders of the English-speaking Canadians were MacDonald and George Brown – Brown, from Alloa, was a newspaper publisher and founder of *The Globe*, Canada's largest national newspaper. In the 1860s MacDonald and Brown united to work in conjunction with the French Canadians to come up with a new self-governing Confederation of Canada that would also include the Maritime Colonies – Nova Scotia and New Brunswick – which led to the foundation on 1 July 1867 and was given the official name of the Dominion of Canada.

By the time of MacDonald's death in 1891, the remaining provinces and territories in British North America had joined the Confederation and become part of the Dominion, with the notable exception of Newfoundland, who did not join until 1949 – although as the rest of Canada will happily point out, the Newfoundlanders have always been a little slow.

British Columbia had only agreed to join the Confederation in 1871 on the condition that there would be a railway that would link Canada from east to west and, after many years of political and financial setbacks – not to mention the practical difficulties faced in laying the line; wild weather, wild country, wild bears, the occasional moderately wild night out (well it is Canada after all) – the Canadian Pacific Railway was completed in 1885. For the first time, the country was connected overland from British Columbia in the west, to Ontario on the east – and so enabling thousands to make a new life without ever having to learn a word of French.

MacDonald was the driving force behind the railway, Kirkcaldy-born Sandford Fleming [95] was the chief surveyor and five Scottish-born businessman formed the Montreal consortium that bankrolled it. Consortium member Donald Alexander Smith, from Forres, had the honour of opening the 2,500 (4,000 km) mile-long line, which included the symbolic driving in of the track's final spike in a ceremony at Craigellachie – named after a village in Speyside – in British Columbia. Normally on such an occasion a commemorative silver or gold spike would be used, but the railway had cost so much to build that no additional expenditure could be countenanced and a standard iron spike was used – which was just as well as Smith's ceremonial blow was so bad that the spike bent completely out of shape and a second one was needed.

In 1887 the first train rolled in to Vancouver, on the Pacific coast, where the settlement's first mayor, Malcolm MacLean, born on the island of Tiree, welcomed it. This signalled the start of Vancouver's growth into a major, international city renowned for its quality of life, sunny climate and beautiful beaches – much like Tiree in fact, although it has often been said that a night out on the Hebridean island is considerably livelier.

Under John A. MacDonald's premiership, the government founded, in 1873, the legendary North-West Mounted Police who would become the Royal Canadian Mounted Police or, as the world knows them, the

Mounties, complete with scarlet jackets and wide-brimmed hats. The official motto of the Mounties is *Maintiens le Droit*, meaning Maintain the Law, but we know them best for the saying, 'The Mounties always get their man', emphasising their image as clean-cut, polite, dogged law-enforcers who have become a proud national symbol of Canada and will travel thousands of miles to ensure that nothing too exciting ever happens.

During the few years that MacDonald was out of office from 1874–78, his replacement was Perthshire-born Alexander Mackenzie – no relation to explorer Alexander Mackenize [44]; thus, for the first twenty-four years as a self-governing nation Canada had a Scot at the helm. One of the most popular patriotic songs of this period was *The Maple Leaf Forever*, written to celebrate Canada's Confederation in 1867 and, although never the official national anthem, the song established the maple leaf as the national symbol, culminating in its inclusion on the Canadian flag in 1965; this new flag replaced Canadian Red Ensign featuring the Union Jack – although there are still some who will argue about whether it is an English- or a French-speaking maple leaf. And, yes, it was a Scot – Alexander Muir from Lanarkshire who immigrated to Toronto as a child – who wrote *The Maple Leaf Forever*.

So, there you have it, pretty much everything

associated with Canada has its origin in Scotland – with the exception of Pamela Anderson who, sadly, is descended from Finnish rather than Scottish Andersons. Canadians are often said to have an indifferent attitude toward their national identity – perhaps not so surprising in view of the Scottish influence – but much in the same way that if you call a Scot English, if you call a Canadian an American, they are likely to really, really lose it, and might even give you a bit of a look.

JAPANESE BOY

49 THOMAS BLAKE GLOVER (1838–1911)

From 1603–1854 Japan followed a policy of isolation from the rest of the world, with one exception, the port of Nagasaki on the south-west island of Kyushi where traders and sailors were free to visit and eat raw fish. But all good things come to an end and in 1854 Japan was forced to open up its economy to the West and, wherever there were new markets and new financial opportunities, you could bet there were Scots.

Henry Dyer from Bellshill in Lanarkshire was the first principal, in 1873, of Japan's first technical college, the Imperial College of Engineering in Tokyo, and Richard Henry Brunton from Aberdeenshire, who worked for the Lighthouse Stevensons, was given the job of building twenty-six lighthouses in Japan from 1868–76. However, preceding both men was Thomas Blake Glover, who was born in the fishing port of Fraserburgh and went to school in Aberdeen.

Foreign travel beckoned for Glover and he found work in Shanghai with the Jardine Matheson trading company before moving to Nagasaki in 1859; here, he sold tea, and probably substances a little stronger, before setting up his own import-export company. Glover's latter business became embroiled in the increasing conflict between the establishment shoguns and the upcoming samurai nationalists and Glover happily sold arms to whoever wanted them. Crucially, however, he built good relations with the samurai clans who would eventually take power in 1868 under new Emperor Mitsuhito.

One of the rebels Glover helped was Ito Hirobumi of the Choshu clan who wanted to study in England. In 1862 Glover helped smuggle him out of Japan by finding him a berth as a deckhand on a Jardine Matheson ship. Hirobumi went to University College, London, and in 1885 became Japan's first prime minister, a leading reformer and one of the most important figures in modern Japanese history.

It is always preferable to be on the winning side and Glover profited handsomely from his political connections. He commissioned the first warships for the new Imperial Japanese Army, with the largest, the *Ryoju Maru*, built in Aberdeen in 1869. He brought the first locomotive to Japan, he owned Japan's first modern coalmine and he built a dry dock in Nagasaki, but by 1870 Glover had over-extended himself and was declared bankrupt. Not that that stopped a wheeler-dealer such as Glover. He became friends with the Iwasaki family who took over his investments and he remained a trusted adviser while the company became the mighty Mitsubishi – the shipping, mining and financial conglomerate at the forefront of the industrialisation of Japan. Meanwhile, Glover was in at the start as a founder of the Japan Brewing Company in 1885 that ultimately became the mighty Kirin brewery. After all, even a Scotsman can drink only so many sakes.

Thomas Blake Glover spent the remainder of his life in Japan. He was a fluent Japanese speaker and in 1863 had a beautiful house built which combined European with Japanese style and overlooked Nagasaki harbour. Glover Garden is today one of the most popular tourist attractions in Japan; it is also known as the Madame Butterfly house because of its association with Puccini's *Madame Butterfly*, the opera based on a short story set in Nagasaki by American author John Luther Long. Long may have based Cho-Cho-San – the character who married an American officer – on Glover's Japanese wife, Tsuru. While there are many references to the opera in Glover Garden, there is no clear link between Glover, his wife and the opera. The Nagasaki authorities may simply have wanted to make Glover Garden more attractive to tourists, a ruse which Glover – ever the opportunist – would no doubt have heartily approved of.

CAN'T GET YOU OUT OF BARRHEAD

50 ANDREW FISHER (1862–1928)

When you think of Australia the song that most readily springs to mind is probably *Waltzing Matilda* – or, perhaps, the *Neighbours* theme tune. The lyric for *Waltzing Matilda* was written in 1895 by poet and author Andrew Barton Paterson, who was the son of a Lanarkshire emigrant and wrote under the pseudonym, The Banjo; the song's tune derives from an old Scottish song, *Thou Bonnie Wood o' Craigielea*, by James Barr from Ayrshire. In 1903 Angus-born James Inglis, whose Australian company sold Billy Tea, included the sheet music for Paterson's song with every packet of tea – the perfect cuppa for jolly swagmen everywhere.

For all the enduring fame of *Waltzing Matilda* down under, it has surprisingly never been Australia's official national anthem. *God Save the Queen/King* was the national anthem until 1977, when it was replaced by *Advance Australia Fair* – written in 1878 by Port Glasgow-born Peter Dodds McCormick, who had immigrated to Sydney where he became a teacher. Since 1977 there has been much criticism that *Advance Australia Fair* is not especially memorable or catchy as a national anthem and it would have been better if they had gone for the more populist *Waltzing Matilda* instead – even if all the references to sheep-stealing and drowning might not be entirely appropriate when you are one of the leading economic nations in the world.

Emigration from Scotland to Australia remained steady throughout the 19th and 20th centuries and, during the latter, New South Wales, followed by South and Western Australia were the destinations of choice for Scots. There was a surge in emigration in the wake of World War II and the Australian Government, keen to increase the population as long as said immigrants were white, followed a policy from 1945 to 1972 that any Brit who could afford £10 would be welcome and, what's more, the Aussies would pay their travel and resettlement costs. Such a financially favourable offer was particularly attractive to Scots and today over 130,000 Australian residents were born in Scotland, and most of those came from the era of the Ten Pound Poms, although being Scots most had kept their receipt just in case they ever decided they wanted to go back.

More than 1.5 million Australians claim Scottish descent, although the true number may be considerably higher. And wherever you find large numbers of Scots you will always find Scots wanting to take over, except in Scotland where such an idea is clearly preposterous. When in 1901 the various self-governing colonies finally came together to form the Commonwealth of Australia with a federal parliament, George Reid from Renfrew was elected leader of the opposition and in 1904 became, briefly, prime minister. Andrew Fisher, however, the second Scot to become prime minister made an even more lasting impact on the new country.

There are many parallels in the early lives of Andrew Fisher from East Ayrshire and Keir Hardie [10] from North Lanarkshire: both were working in the coalmines by the time they were thirteen; neither had any formal schooling and both taught themselves to read and write; both were involved with miners' unions; both lost their jobs and were blacklisted by mine owners and both ended up playing a central role in politics for the Labour Party. The only major difference was that while Keir Hardie began his career as a journalist in Scotland, Andrew Fisher chose to emigrate and in his search for work arrived in Queensland in 1885.

Two years before Keir Hardie founded the

Independent Labour Party in 1893, the first ever meeting of the Labour Electoral League – the precursor to the Australian Labour Party – took place in the Unity Hall Hotel in Balmain, Queensland. Now a thriving district of Sydney, Balmain takes its name from Perthshire-born surgeon William Balmain. Andrew Fisher became involved in Labour politics in Queensland during that decade until, in 1899, he was appointed a Queensland government minister in the first Labour government anywhere in the world, although it would only be in office for one week before, inevitably, running out of money.

Two years later in 1901, when the Commonwealth of Australia came into being, Fisher became a Federal MP; he became leader of the Labour Party in 1907 and prime minister of a minority government between 1908–09 before returning to power in 1910 when Labour had their first parliamentary majority. And it was this 1910–13 administration that proved one of the most important in recent Australian history with: the introduction of pensions and increased workers' rights, the creation of the Australian Navy, the founding of the State Bank of Australia, now as the Commonwealth Bank one of the world's largest. In addition, this period saw construction begin on the Trans Australian Railway, which on completion in 1917 connected the country from east to west; the line included a 300-mile (483-km) straight, the longest straight railway line ever laid in the world – and bugger any sacred aboriginal land that might be in its way.

It was also Andrew Fisher who plumped for Canberra as the location for the new capital of Australia and laid the foundation stone there in 1913. It had been one of the stipulations of Federation that to avoid having to choose between the two great rivals, Sydney and Melbourne, the capital should be built from scratch and sited roughly halfway between the two, and the Canberra district fitted the bill; Canberra is derived from a native Australian word meaning, meeting place. The brand new city finally replaced the temporary capital of Melbourne in 1927 and over subsequent years has indeed proved successful in bringing Melbourne and Sydney together, if only in complete agreement about how dull Canberra is.

No sooner than the second Fisher Labour government ended than he returned for a third, and final, time in September 1914. By this time he had changed the name of the Australian Labour Party to the Australian Labor Party in those blissfully innocent days when international socialists hoped that America might one day turn pinko-commie as well.

Two months later, World War I had begun and Fisher

pledged, despite his personal reservations, Australian support to King and Empire. Enthusiastic Australian volunteers joined up and were shipped to Egypt where they awaited further instructions. The original plan had been for the Australians and New Zealand Army Corps (ANZACs) to join the Allies on the Western Front but the British decided they should instead be diverted to an all out attack on Turkey, a German ally, and on 25 April 1915 the Anzacs duly landed on the Gallipoli peninsula. Even by World War I standards the campaign was disastrous, the Anzacs were sent to the wrong landing points on day one and the whole thing went downhill from then on.

On the beaches of Turkey, 80,000 Turks, 20,000 British, 10,000 French and 2,700 New Zealanders died, and for Australia, which had never before been involved in major armed conflict, the impact of 8,700 dead and 19,000 wounded would have a devastating impact. When Fisher belatedly discovered the extent of the casualties and the dreadful conditions that those remaining were having to endure under British leadership, despite the fact that he had already determined to resign the premiership, Fisher used all his influence to ensure that the decision to evacuate the Allied forces, including the Australians, and abandon Gallipoli had been made and that the bloodbath would belatedly come to an end.

Andrew Fisher left office in October 1915 and, succeeding George Reid, became Australia's second High Commissioner to Britain. ANZAC Day is commemorated annually on 25 April in honour of those who lost their lives in the Gallipoli campaign. It has become part of the received wisdom of Australian history that as a consequence of Gallipoli Australia lost its innocence and went from being a colony beholden and dependent on a distant empire who sent the young men of Australia to their death for no good reason and instead became an independent nation state in its own right with its own history. Granted, a white only history for a white only country, but then again nobody's perfect.

Practical Scots

Scottish inventors and engineers have a lot to answer for. Today there are undoubtedly many Scottish men who are fantastic cooks, wonderful lovers and erudite and sparkling conversationalists, as well as many Scottish men whose idea of DIY is knowing where the Yellow Pages is so that they 'can get a man in'. However thanks to centuries of amazing Scottish inventors and inventions it is for their practical, rather than their more cerebral side, that Scots are known throughout the world. And this reputation has become entrenched even at home, as no matter how delicious your foie gras, how informed your discourse on the merits of the Swedish *Wallander* TV series compared to the British version and how creative your foreplay might be, it is not going to get you very far if you still haven't, as promised, got around to putting those shelves up. (And talking of making the Earth move, was it not James Porteous from Haddington who, in 1883, invented the Fresno Scraper – one of the world's most important earth-moving machines?)

I FEEL A LITTLE FLUSHED

51 ALEXANDER CUMMINGS (1733–1814)

Toilet humour has a long and pungent history in Britain and it is therefore highly appropriate that the man who is often credited with being the inventor of the modern lavatory was a 19th-century Yorkshire plumber by the name of Thomas Crapper. Crapper invented the floating ball cock, and he did much to *popularise* the domestic flush lavatory but he was *not* its inventor: it is a Scotsman who has that particular honour.

Alexander Cummings was born in Edinburgh in 1733 and worked as a watchmaker in London. Perhaps it was his upbringing that heightened his interest in all things sanitary, as Scotland's capital city in the 18th century was one of narrow streets and closes packed with thousands of people and overcrowded tenements up to fourteen stories high. Sanitation consisted of chamber pots, the contents of which would be discarded out the window from a great height with the warning call, 'Guardyloo!' – from the French *guardez l'eau* meaning 'beware the water'. If you were especially unfortunate or were not especially popular with the thrower there was always the possibility of a direct hit, but even the most agile of pedestrians would also have to be aware of the constant danger of splash back – although, on the positive side dry cleaning companies were doing very well.

There had been attempts over the centuries to design a practical toilet or water closet for use in the house, but all had foundered at the odour-test stage and up until the late-18th century it was still generally believed that it was better out than in. In 1775 Cummings patented a flush toilet that had both a water trap of standing water below the bowl and a valve in the form of a S-bend that would ensure that there was some water in the toilet bowl before use and, once your business was done and you had ensured that the toilet lid remained in the upright position, additional water was added to clean and refill the bowl, so reducing unpleasant aromas.

It would take another hundred years and several redesigns for the flush toilet to begin to be commercially produced and it would not be until the 20th century that the 'little boy's room' would become a domestic feature whereas previously little boys had been sent outside. The toilet that we know and love, and occasionally clean, today still retains much of the basic premise of Cummings' design and has transformed the world that we live in, although having to pay more than 20p for using one is just taking things too far.

ABSOLUTELY STEAMING

52 JAMES WATT (1736–1819)

The Industrial Revolution was born in Greenock. Well, technically the period of history in the 18th and 19th centuries that transformed the economies of Britain, Germany, France, the US and Belgium from being predominantly agricultural to predominantly industrial began in Glasgow in 1765 but, much like Belgium, Greenock does tend to get neglected when discussing major historical events, so it is always nice to give it a mention when you get the chance.

James Watt was born in Greenock in 1736 and worked at Glasgow University as an instrument-maker, and it was there that he invented a condensing steam engine. Both steam and machines that utilised steam had been around for centuries and by this time the most developed steam engine had been invented by the Englishman Thomas Newcomen in 1712. Newcomen's engine was used for pumping water out of mines and it was a broken one of these that the University had shipped to London for repairs. Watt arranged for the engine to be shipped back to Glasgow in 1763 in order to have a chance to see, close up, how the engine worked and he realised that he could make it far more efficient by having a separate condenser where the steam could be kept cool, while the steam in the other chamber would be kept permanently hot. This new cycle required far less fuel and would increase the engine's speed resulting in greater power being produced and his redesigned steam engine received its patent in 1769.

Watt had produced a prototype of his new design and the next step was to raise the funds to produce working machines. An appearance on *Dragon's Den* proved unsuccessful as the dragons as per usual didn't understand the concept, although at least there was the satisfaction of Peter Jones insisting on touching the hot chamber, and it was not until 1775 with Watt's partnership with Matthew Boulton in Birmingham that his new, improved steam engines began to be built.

In 1782 Watt invented a double-action engine, that further improved the efficiency and power of the engine and to help promote his invention he came up with the term 'horsepower' to explain to potential customers the benefits of his engine compared to the standard beasts of burden. Watt calculated that a horse could turn a mill wheel with a radius of 12 feet (3.7 m) 2.4 times a minute with a force of 180 lbs (82 kg) and this was the equivalent of one horsepower – and the term horsepower remains to this day an accepted unit for measuring engine output throughout the world. Watt's original steam engine had 10 horsepower and his double-action engine would have 15 horsepower and by 1800 1,000 of his engines were at work in factories and coalmines around Britain. The Industrial Revolution had well and truly begun and horses were not invited. Millions would leave Britain's rural areas to work in factories, where on the one hand you did not have to stop work due to rain or snow or nightfall, but on the other hand you never stopped working.

The massive impact James Watt had on the worlds of technology, engineering, industry and *Top Gear* was recognised when the watt (W) was adopted as the SI (international system) unit for power, and is to be found in every household in the form of electrical power. The term for 1000 watts is a kilowatt (kW) and the term for 1,000,000 watts is a megawatt (MW), but for many people watts are associated with domestic light bulbs that are commonly found as 40 or 60 watts in recognition of the number of people from Greenock it takes to change one.

BANG, BANG, YOU'RE DEID

53 PATRICK FERGUSON (1744–80)

Just because John Paul Jones [91], James Wilson [41] and many others born in Scotland played such a prominent part in the American War of Independence do not think that all the Scottish emigrants to America were all on the rebel side. The Scots-Irish immigrants – from Northern Ireland – were overwhelmingly in the American camp, but the Scots who throughout their history had made disunity into an art form were not going to make an exception just because they were in a different continent several thousand miles away from home. Even Flora MacDonald, the Jacobite heroine imprisoned by the British for her role in smuggling Bonnie Prince Charlie over to Skye in 1746, became embroiled. Flora had immigrated to North Carolina in 1773 and when war broke out she found herself, as did many Highlanders, on the side of the Crown. Flora returned to Skye in 1783 and the remaining years of her life were relatively free of incident, although with her having been on the losing side twice in two different countries, nobody wanted to be in Flora's team at bridge nights.

The other Scots who fought on the losing side were those who served in the British Army and the most famous of those was Patrick Ferguson from Edinburgh. Ferguson came from an eminent Scottish family and served in the British army as an officer. He was an excellent marksman and became interested in improving the standard front-loading musket that was then used by soldiers. Ferguson adapted previous designs for breech-loading rifles, with the breech being at the rear of the gun barrel, and patented his new rifle in 1776. The advantage that his rifle had over the musket was that it was lighter, could fire more rounds per minute and, crucially, could be fired and reloaded from a prone position, as compared to the musket where you had to be standing up.

When war broke out more than a hundred of Ferguson's rifles were produced, the first breech-loading rifles ever to be used in warfare, and Ferguson was sent to America to lead a group of riflemen. Ferguson and his rifles proved decisive in many early engagements and he was promoted to major, but in 1780 he was defeated and killed at the Battle of King's Mountain in South Carolina – a decisive turning point for the British, although the Americans magnanimously didn't change the name of the mountain.

A famous story concerning Ferguson and his rifle was in 1777 when he was scouting before the Battle of Brandywine and spotted an American officer on horseback. His uniform indicated that the officer was of high rank and he was well within range of fire, but the officer had his back to Ferguson and he therefore was not prepared to shoot. Later Ferguson was informed that the officer in question had been George Washington, but he did not regret his decision to shoot the future president even though by doing so he could have changed the outcome of the war – it would not have been the honourable course of action to take and furthermore, the other marksman behind the grassy knoll was in a far better position.

Ferguson's death meant the end of the Ferguson rifle, and breech-loading rifles went through many further designs before finally replacing front-loaders as the military's firearm of choice. In 1879 a Hawick-born gunsmith and inventor by the name of James Paris Lee patented the first successful rifle with a detachable magazine box of ammunition – ideal for rapid firing and reloading. Lee's designs were adopted by different countries and his Lee-Enfield, bolt-action, magazine-fed repeating rifle was the gun used by the British Army and other Commonwealth armies from 1895, through both World Wars until the 1960s; and today, more than a hundred years since it was first produced, it is still used by the police in India. Or to paraphrase the catchphrase of another illustrious son of Hawick, the late, great, rugby union commentator Bill McLaren, 'They'll be shooting in the streets of Bangalore tonight'.

JEANIE BY GASLIGHT

54 WILLIAM MURDOCH (1754–1839)

Of all the inventions that Scots have given to the world, the one that got away was the locomotive. True, the train could not have existed without the steam engine of James Watt [52], but it was an Englishman, Richard Trevithick from Cornwall, who built the first working steam locomotive in 1804. And although the honour could, and perhaps should have belonged to a Scotsman, considering how many other inventions the Scots have claimed credit for, sometimes rather dubiously, we should perhaps not be too greedy when the occasional one passes us by.

William Murdoch came from Auchinleck in Ayrshire and from an early age was one of those boys who was obsessed with science and inventions rather than playing football and pulling girls' hair. The Soho Foundry in Birmingham run by Watt and Matthew Boulton was the epicentre of engineering innovation, and in 1773 Murdoch made the 300-mile (483-km) journey on foot and asked for a job. Impressed by both his determination and his credentials, Watt and Boulton sent him to Cornwall to help install their engines in the tin mines – although thankfully he did not have to walk there.

Murdoch was responsible along with Watt for

improvements to the steam engine and he began to develop the idea of producing an engine that could power a vehicle. In 1784 Murdoch produced a working model of a steam carriage and two years later gave a public demonstration in Cornwall. Watt however was becoming less than impressed about the amount of time his employee was spending on this secondary-project and was concerned about the dangers of using high-pressure steam on a moving vehicle and he told Murdoch in no uncertain terms to desist. Murdoch reluctantly had to do what his boss said, although in his spare moments he would still spend his time in his shed tinkering with his models until one dark evening in Redruth, Cornwall, he took his updated steam carriage for a drive, frightening the life out of the local vicar who thought he had met the devil himself. Although what exactly the vicar was doing out in the middle of the night has never been fully explained

Also living in Redruth was one Richard Trevithick, who also worked on steam engines in the mining industry and Murdoch was more than happy to show his neighbour his steam carriage. In 1801 Trevithick built his own version and in 1804 in South Wales demonstrated the first journey in the world by a steam locomotive on a railway when ten tons of iron were pulled ten miles (16 km) in just over four hours. News of this momentous event would bring back bad memories for the Redruth vicar, while Murdoch was said to be so shocked that he almost choked on his Cornish pasty.

The train may have been born in Cornwall, but Scots did play an integral role in steam-driven advances. In the 19th century Glasgow, and especially the city's Springburn district, was one of the largest producers of locomotives on the planet and it was George Turnbull from near Perth who, as chief engineer of the East Indian Railway Company, played a crucial role developing the rail network in India. Between 1851-62, Turnbull was tasked by the Governor-General, James Broun Ramsay, Marquess of Dalhousie, from Dalkeith in Midlothian, to lay the 540-mile (869-km) railway line connecting India's largest city and then capital, Calcutta, inland to the holy city of Benares. This line would then continue on to Delhi, and eventually to cover the entire continent – and everybody would still be home in time for tiffin.

Experiments with steam-power in boats began in the 1730s, and the most ambitious of these by engineer William Symington who was born in Leadhills in South Lanarkshire. Symington had worked on various prototypes for twenty years until in January 1803 he launched the *Charlotte Dundas* on the Forth and Clyde Canal. In as much as it had a purpose, it was the world's first 'working' paddle steamer and successfully hauled two, seventy-ton barges over a distance of twenty miles. But due to concerns over potential damage to the canal further sailings were banned and a distraught Symington lost his funding. Four years later, one of James Watt's engines was used in the world's first commercial steamboat service, launched by American Robert Fulton in 1807 to operate on the Hudson between New York and Albany.

Back in Cornwall in the 1790s, William Murdoch came up with another invention that would change the world. Investigating the properties of hot coal, he had heated coal in a kettle with a thimble with holes in it fixed over the spout and discovered that the gas escaping through the holes, when ignited, made light. In 1792, after first making sure that the vicar was on holiday, he produced enough gas to light his entire house. Six years later he was able to light up the foundry works of his employers in Birmingham and in 1802 he gave his first public demonstration of gas lighting to an astonished audience, before the dawning realisation that the added illumination only highlighted the fact that most of the observers could do with a good wash.

Before gas lighting, towns had been lit by oil lamps, torches and candles, but coal gas was more efficient to produce and gas lamps appeared on the streets of London as early as 1807, the first gas company was founded in 1809 and the first gas lighting in America came in 1816. Soon every town, city and factory in Europe and America used gas lighting and going out for the evening became a far less hazardous proposition – although if it was a hazardous proposition that you were after, then all this outdoor lighting made keeping such liaisons discreet increasingly difficult.

For William Murdoch, the faithful company employee, his invention did not make his fortune but James Watt, while keeping the profits for himself, did at least reward Murdoch for all his hard work with a generous bonus, which being the generous man that he was, he decided to share with the still recovering vicar of Redruth.

Gas lights would eventually be replaced by electric lights and it is generally accepted that the inventor of the electric lighting was American Thomas Edison, who demonstrated his light bulb in New Jersey in 1879, with Edison's middle name of Alva after the Clackmannanshire town revealing a distant Scottish ancestry. However, there have been other claimants with one of the most intriguing being James Bowman Lindsay, a teacher born near Arbroath. In 1835, he demonstrated at a series of public meetings in Dundee – which were covered by the local press – an electric light bright enough to read a book or newspaper by from a distance of one and a half feet. Sadly, as Lindsay never patented or published a record of his electric light, we do not know what type of apparatus it was, how it worked or what happened to it, or for that matter what the inventor was reading, which means that we will never know for certain if the first demonstration in the world of electric lighting was accompanied by a Dundonian exclaiming 'eh saw the light'.

ALL ROADS LEAD TO CUMNOCK

55 JOHN LOUDON MCADAM (1756–1836)

It was the Romans who brought stone roads to Britain and once they had left in the 5th century the budget for maintaining the transport infrastructure was severely cut back with roads becoming mostly soil-based dirt tracks. In Scotland, where said Romans had restricted themselves to practising their wall-making skills, the roads were particularly bad. Not until the 18th century did the British army begin to build roads into the Highlands, all the better for their troops to maintain law and order, but by the end of the century horse and carriage insurance premiums were at an all time high and even chickens were increasingly reluctant to cross the road due to the numerous pot holes they would have to traverse before getting to the other side.

When the Ayr-born merchant John Loudon McAdam,

who had made his money in New York, returned to run an estate in Ayrshire he was appalled at the state of the roads he had inherited and decided that something must be done. McAdam began to experiment with different forms of road construction to see what improvements he could make. He tried taking the high road and he tried taking the low road, and found that if the level of the road was raised the drainage was much improved. McAdam then perfected a new form of construction that involved putting down two layers of equally sized crushed stones as a base and on top an additional layer of smaller stones, also of equal size, bound together with gravel, with the layers then flattened by using a heavy roller. This created a smooth hard surface and revolutionised road construction around the world.

In 1815 McAdam was appointed surveyor-general of the roads in Bristol and so successful were the improvements he made that by 1827 he was appointed surveyor-general for all of Britain's metropolitan roads. McAdam's improvements became known as 'macadamisation' and were taken up in the US and throughout Europe.

By the end of the 19th century, with the introduction of motorised vehicles, McAdam's roads were becoming increasingly liable to dust, and in 1901 Edgar Purnell Hooley from Wales patented the use of an additional layer of tar that would then be flattened with a steamroller. This reduced the amount of dust and became known as tarmacadam or Tarmac for short. In the 20th century the original macadamisation process, that involved a large amount of manual labour, would be replaced with the use of asphalt, but the term Tarmac has been retained as a generic term for any road surface.

McAdam continued to work right through his seventies and made regular trips back to Scotland where he died aged eighty at Moffat, although thanks to his innovations his final journey was much faster and more comfortable than those of his youth.

OTTERS CONCERNED BY NEW BRIDGE PLAN

56 THOMAS TELFORD (1757–1834)

The name of John Loudon McAdam [55] is often bracketed with that of Thomas Telford. Telford was born near Langholm in Dumfries and Galloway and was the best-known engineer of his day. He made his name in Shropshire, where the new town of Telford would later be named in his honour, and in his career he was responsible for the construction of more than 1,000 bridges, 900 miles (1,448 km) of road and numerous canals as well as the occasional conservatory for friends and family.

Before the advent of railways, canals were the transport network of the Industrial Revolution and Telford designed the Caledonian Canal between Inverness and Fort William and the Göta Canal that crossed the south of Sweden from coast to coast linking the two largest cities of Gothenburg and Stockholm. Telford's commission to design the Göta Canal was no coincidence, as ever since the city's founding in 1621 there had been strong trade links between Scotland and Gothenburg and a prominent Scottish community keen to sample the local meatballs, although Telford had to explain to a disappointed Swedish king that a flat-pack canal would not be possible.

Telford was responsible for the building of the 286-mile (460-km) new road between London and Holyhead in North Wales and the construction of, at the time, the longest suspension bridge in the world, the Menai Suspension Bridge, between the mainland and the island of Anglesey, which was completed in 1826. The Menai Bridge is Telford's most famous construction and has gained almost mythical status, partly due to Anglesey's long history of being the last stronghold of the druids, but mostly because as hardly anyone ever goes to North Wales very few people have actually seen it. Telford was given the nickname the Colossus of Roads, although considering the original Colossus partly collapsed because of an earthquake, this was perhaps not the best moniker for someone who built bridges for a living.

Even more famous than the Menai Bridge is that icon of the Scottish painting and decorating industry, the Forth Bridge. Completed in 1890 it was the longest bridge in the world and Scottish engineer William Arroll from Houston in Renfrewshire supervised its construction. Not satisfied with building one national icon, Arroll's firm was also involved in the construction of one of the most recognisable landmarks in London – Tower Bridge. It is a bascule bridge, which means it could be raised in the middle to allow ships through, was completed in 1894, and became such a recognised landmark that it was often mistakenly called London Bridge, which was in fact the next bridge along and had been designed by the 19th-century engineer John Rennie, from East Lothian. When Rennie's London Bridge was sold in 1968 and shipped out to be rebuilt in an Arizona theme park, it was suggested that the American buyers had actually been expecting Tower Bridge, but considering that there was no river in their theme park, having the wrong bridge was the least of their concerns.

William Roy from Carluke in Lanarkshire was an engineer who may not have built any roads or canals, but without his contribution those who did would have found life considerably more difficult. Roy worked for the British Army as a military surveyor and in 1783 began the epic task of accurately mapping the whole of the British Isles using the process of triangulation, making Britain the second country after France to have been mapped. The process was not completed until 1853, partially due to numerous unfortunate surveyors disappearing into the peat bogs of Ireland, and Roy, who died in 1790, would not live to see its conclusion.

Roy would also sadly not live long enough to see the official founding in 1791 of the Ordnance Survey, the government mapping agency that he more than anyone had helped create. The Ordnance Survey continues to this day to be an organisation known internationally for the quality and accuracy of its mapping, and an Ordnance Survey map is, along with a Snickers bag, one of the two items that no hill walker can do without.

No discussion of great Scottish engineers would be complete without mentioning Linlithgow-born Montgomery Scott who for thirty years was chief engineer on the *USS Enterprise*. Scott spent a good part of his early life in Aberdeen, and although in his career he had boldly gone where no man had gone before he never forgot his Scottish roots and once said that at heart he was an 'old Aberdeen pub-crawler' – which also meant that he had already encountered most forms of life to be found in the universe. When *Star Trek* was first commissioned in 1966 and Canadian actor James Doohan was given the role, the character of Scotty was not at that point Scottish, but as Doohan said, 'All the best engineers in the world come from Scotland' – which even Leonard Nimoy had to accept was perfectly logical.

THE HAMMER OF THE SCOTS II: THE RISE OF THE MACHINES

57 JAMES NASMYTH (1808–90)

If James Watt [52] powered the Industrial Revolution and William Murdoch [54] lit the Industrial Revolution, then Glasgow and the West of Scotland was the engine room for the Industrial Revolution. Exploiting the natural, abundant supply of coal and iron-ore in the Lowlands, and a growing population who were willing, inexpensive and would work ridiculously long, intensive hours down the mines and in factories in the name of their protestant work ethic – even if increasingly they happened to be Catholic – Clydeside became synonymous in the 19th century with heavy industry and in particular with the railways and with shipbuilding; for a time, the Clyde was building twenty percent of all the ships in the world. By the end of the century the city's population had topped one million and Glasgow was calling itself the Second City of the British Empire, thereby articulating a long-held pride in its superiority toward the likes of Birmingham, Bombay and Calcutta, while not for one second contemplating that rather than being the second city of someone else's empire they could actually become the first city of their own nation.

Three of the four founders of the first successful transatlantic commercial shipping company, the British and North American Royal Mail Steam Packet Company that would later become the Cunard Line, were Scottish and the company's first steam ocean liner to provide a regular passenger and cargo service was *RMS Britannia*, which was built in Greenock and began sailing from Liverpool to Boston in 1840. Since the paddle-steamer *Comet*, built at Port Glasgow in 1812, operated the first commercially successful steamship service in Europe, through to the some of the biggest ships in the world of their time – *RMS Lusitania* in 1907, *RMS Queen Mary* in 1934 and *RMS Queen Elizabeth* launched in 1938 – Clydeside's shipyards thrived and unlike the city of Glasgow were second to none.

But had innovative engineers not paved the way, such ships could have been neither designed nor built. In 1828 James Beaumont Neilson, from Glasgow, invented and patented a new process for smelting iron in which the air was preheated before entering the blast furnace. Known as 'hot blast', it needed less coal to keep the furnace hot and, therefore, readily available coal, rather than coke or charcoal, could be used. The furnaces may still have resembled the fires of Hell but, thanks to Neilson, even Lucifer had to admit they were much more fuel-efficient.

The invention of the steam hammer was to the 19th century what the invention of Watt's steam engine had been to the 18th century. James Nasmyth was born in Edinburgh, the son of noted landscape artist Alexander Nasmyth. Alexander, however, had also long been interested in mechanics and his son inherited this passion and studied to become a mechanical engineer. James Nasmyth moved to London to find employment before setting up his own foundry in Lancashire in 1836 building large machine tools alongside locomotives. In 1838 he was approached by Isambard Kingdom Brunel's Great Western Steamship Company, which was building the *SS Great Britain,* the then largest ship in the world. Brunel was in the market for a yet to be invented forge hammer powerful enough to make the ship's paddle shaft and hoped Nasmyth would solve the problem. Rather than a larger version of the traditional manual hammer, Naysmith designed a steam-powered version that would give both power and precision. However, by the time he had completed the design process in 1839 Brunel had completely changed his mind and decided to use screw propellers rather than paddle wheels, dashing Nasmyth's hopes of building his new invention.

Three years later, while visiting foundries and ironworks in France Nasmyth popped in to the Le Creusot ironworks in Burgundy, owned by the Schneider brothers, and there to his great surprise was a fully operational steam hammer that looked exactly like the one he had designed. Zut alors! In Nasmyth's absence, one of the Schneiders had visited his Lancashire foundry and been shown the designs which he had then copied. As neither party had ever bothered with a patent, Nasmyth hot-footed it back to England where – to the consternation of the French – he patented and built his steam hammer and with great panache demonstrated publicly how his machine could provide great force yet was so precise it could crack an egg inside a wine glass without breaking the glass; and to further infuriate the French he drank two glasses of Burgundy and repeated the demonstration with identical results.

James Nasmyth had become the first to build a machine that could, in turn, build another machine. With the introduction of the steam hammer came the second stage of the Industrial Revolution; machines became faster, bigger and stronger – and less likely to ask for holidays. Nasmyth proceeded to build steam hammers and sell them throughout the world, and in 1843 he came up with a steam-powered pile driver that reduced the time taken to do the job manually by many, many hours. He retired in 1856 at the age of only forty-eight and spent the remainder of his life happily enjoying the wealth that the steam hammer and pile driver had brought him, and no doubt raising an occasional glass of best French wine in the process.

WHAT IS IT WITH SCOTSMEN AND RUBBER?

58 JOHN BOYD DUNLOP (1840–1921)

Sometimes great minds think alike and come up with the same idea. In 1887 a vet from North Ayrshire, John Boyd Dunlop, who had settled with his family in Belfast, was concerned that the solid rubber tyres on his son's new tricycle were pretty unforgiving and his son was having a far from comfortable ride on the city's cobbled streets. Being a doting Scottish father, Dunlop decided that something must be done to alleviate his son's discomfort and came up with the idea of taking several lengths of rubber garden hosepipe, wrapping them around the tricycle wheels, gluing them together and then inflating them with air. His new homemade tyres made the tricycle not only more comfortable to ride, but much to the concern of Mrs Dunlop, also considerably faster. Buoyed by the success of his home-made pneumatic tyres, Dunlop put them to the test in a Belfast bicycle race where everyone was astonished that his bike was so much faster than of all the others – even taking into account the winning cyclist's failed drugs test.

Realising he was on to a winner, Dunlop patented his pneumatic tyre in 1888. The following year he formed the Dunlop Pneumatic Tyre Company in Dublin and with the advent of the bicycle craze decade that was the 1890s found himself ideally placed to produce inflatable rubber tyres for the international market. Dunlop and his tyres were poised to take over the world, or so he thought. His patent, he discovered, was invalid. Unbeknown to Dunlop, almost forty years earlier, another Scot, Robert William Thomson, had come up with pretty much the same idea and even more annoyingly had taken out a perfectly legitimate patent – although on the positive side he was dead and therefore couldn't sue.

Thomson was born in Stonehaven and worked as an engineer throughout Scotland before moving to England, then Indonesia, and then back to Scotland. He was a serial inventor and possessed dozens of patents with his greatest success overall being a road steamer, or steam traction engine. Designed in 1867, his road steamer had solid rubber tyres and was intended to carry huge loads by road, thus negating the need for rail tracks or horses. By the time of his death in 1875, Thomson's steamers were being used around the world and had made him a wealthy man: as for his invention that would ultimately change the world he had long since abandoned and forgotten about it.

By 1845 the vulcanisation of rubber had become logical and Thomson was one of many who were experimenting with it. He came up with a hollow rubber belt inflated with air inside a leather tyre that could be attached to the wheel of a horse-drawn carriage. He called his invention 'aerial wheels' and in 1847 demonstrated it successfully at Regent's Park, London. Thomson had patented his invention in 1845, but with rubber an expensive commodity and with no obvious market for his 'aerial wheels' at the time, he moved on to different projects; accustomed as he was to patenting so many of his inventions there was always the chance that the occasional one might come to nothing.

News of Thomson's tyre patent had come as a blow to the Dunlop Company as it ruled out exclusivity in the market, but with the growth of first the bicycle and then the automobile industry there was more than enough demand to keep everyone happy. The Dunlop Company began producing tyres in England in 1891 and would eventually establish its headquarters at a giant factory, Fort Dunlop, in Birmingham and became one of the largest tyre manufacturers in the world and the dominant producer in Britain until 1985 when the company was sold and broken up.

John Boyd Dunlop, however, having agreed to being bought out of the Dunlop Tyre Company in 1896, missed out on the huge wealth and profit that would have come his way, and lived a quiet, but comfortable, life in Dublin, with Mrs Dunlop always making sure that a safety helmet was worn at all times.

ARBROATH SMOKIE ENGINE

59 DAVID DUNBAR BUICK (1854–1929)

On the basis of the Scots' contribution to roads, tyres and bicycles – courtesy of John Loudon McAdam [55], Robert Thomson and John Boyd Dunlop [58] and Kirpatrick Macmillan [63] respectively – you might expect them to, also, be integral to the invention and development of the automobile. True, we should not forget William Murdoch's [54] steam carriage of 1786 and we should also bear in mind Aberdeen-born Robert Davidson's four-wheeled, four-mile-an-hour, battery-driven machine – the first known electric locomotive – built in 1837 and tested successfully on the Edinburgh to Glasgow railway line in 1842. But as far as the invention of the modern automobile was concerned in the 1880s and '90s it was the Germans and the Americans who were at the forefront – although Scots would still gain an honourable mention.

Alexander Malcolmson from Ayrshire could have – but failed to – make millions in the car industry. Having emigrated to Detroit and getting rich in the Michigan coal industry, Malcolmson in 1902 agreed to invest $7,000 and form a partnership with a local man and up-and-coming designer by the name of Henry Ford and, with that, the Ford Motor Company was founded the following year. Its first car was the Model A and by 1906 the company had reached the Model K. But Malcolmson was convinced that the future lay in luxury, rather than mass-market, cars and in one of the less prudent decisions in the history of modern commerce he agreed to sell his shares. Granted, he garnered a tidy profit of $175,000, but two years later in 1908 Henry Ford introduced the Model T which set Ford on his way to becoming one of the world's first billionaires, and it is therefore not too surprising that subsequently when on the road Malcolmson made sure that he never allowed a Ford to cut in.

However, Alexander Winton from Grangemouth is said to be the man who started up the first car-manufacturing business in the US. Winton immigrated to Cleveland, Ohio in 1896 and set up a bicycle company before diversifying into motorcar production. The following year he founded the Winton Motor Carriage Company, sold his first car in 1898 and by the end of the century was producing more than a hundred a year. By 1924 Winton could no longer compete with the mass-production market and he stopped producing cars, but continued to work improving engines and car designs until his death in 1932.

In terms of iconic marques America, of course, has Ford, but the classic Buick was named after a Scot. David Dunbar Buick was born in Arbroath but emigrated with his family at the age of only two and grew up, as did Malcolmson, in the future motor town of Detroit. Buick worked for a plumbing company and showed his early genius as an inventor by coming up with an enamelling process that enabled cast iron bath tubs to be white, and which made him enough money to devote time to a new passion – building his own motor car.

In 1902 he originated the revolutionary, overhead valve engine which would ultimately give cars more power than ever before and in 1902 formed the Buick Manufacturing Company and then started producing his own cars in 1903, one month before Ford. It was soon clear that although Buick was an inspired designer, he was not a great businessman, and as the debts built up he was obliged to bring in investors to keep the company afloat. Production at Buick Motors began slowly to increase, but although David Dunbar Buick was officially company president he had little say in the running of the business he had started and, in 1906, disillusioned and disheartened, he sold his remaining shares for a nominal sum.

One of the investors brought in to rescue Buick was a William C. Durand who, like Henry Ford, realised that mass production was the way to go. By 1907 Durant was producing and selling more than 4,000 Buicks a year, more than any other car in America, and on the back of this success founded General Motors in 1908. By the 1930s General Motors had overtaken Ford to become the largest automobile company in the world, and remained so for a further six decades.

In its early days the Buick sold to the growing middle classes and drivers looking for leather-clad comfort; during the 1920s sales reached 250,000 a year, rising to 500,000 a year in the '50s with record annual worldwide sales in 1984 touching one million.

And what of David Dunbar Buick, the man who began it all? Buick was a great designer let down by his lack of business acumen. After leaving the company he remained in Detroit and to support his family he worked as a clerk and then as a school inspector right up until his death in 1929 – a forgotten and impoverished figure despite his name being known by millions and the company he helped originate would become the largest in the world.

General Motors did not completely ignore its Scottish founder: in 1937 it introduced above the radiator grille the Buick shield, based on the family coat of arms, which in 1959 became the familiar tri-shield crest we see on every Buick today. Which was all very well, but might have meant a little more if they had done so when Mr Buick himself had still been alive.

NOT ALL ENGINES ARE FEARSOME

60 DUGALD CLERK (1854–1932)

In the early-19th century the technology behind steam engines was yet to advance enough to prevent them exploding from time to time. In 1818 Perthshire minister and part-time inventor Robert Stirling came up with a closed-cycle engine where air or gas is heated and then recycled or regenerated to produce power. The Stirling engine had the great advantages that it was much more fuel-efficient than the steam-based engine and, handily, did not explode. But its disadvantages were that it took longer to warm up and in the early years had a tendency to burn itself out.

When a Stirling engine was used in a Dundee foundry in the 1840s, it broke down so regularly that the Dundonians soon reverted to steam and with the advent of first the internal combustion engine and then the electric motor, the Stirling heat engine was abandoned. However, the Stirling engine has not been completely forgotten: nearly 200 years later, scientists and engineers today continue to work on the technology that may deliver more power for less fuel and, thus, in the 21st century the Rev Robert Stirling's name could, perhaps, become as familiar as that of the other great Scottish inventors.

By the mid-19th century science had embraced the developments of the Industrial Revolution and thermodynamics was all the rage with at its forefront Irish-born, Glasgow-domiciled William Thomson, later Lord Kelvin, alongside Edinburgh-born William Maquorn Rankine who was professor of engineering at Glasgow University. Kelvin and Rankine both produced comprehensive studies of the science behind heat engines, which included those driven by steam. Rankine wrote numerous manuals from 1859 onward that became standard textbooks for budding engineers around the world. He also produced a temperature scale called, unsurprisingly, the Rankine scale, which is still used by engineers today, but in the Fab Four of scales

– Celsius, Fahrenheit, Kelvin and Rankine – the Rankine would clearly be the Ringo of the group.

Dugald Clerk was a Glaswegian and he studied engineering at what is now Strathclyde University. In 1876 a German, Nikolaus Otto, developed the first practical four-stroke internal combustion engine and in doing so revolutionised the world of engineering and machines. The young Clerk was hugely impressed both by Otto's engine and by the other pioneers of this new technology, but became convinced that it must be possible to achieve a workable two-stroke engine. Clerk's two-cycle engine had two equally sized cylinders and was gas-fuelled; it took him four years to develop and in 1881 he was the first to patent a two-stroke internal combustion engine.

The new engine immediately attracted interest and Lord Kelvin installed a Clerk cycle engine to run electric lighting at his Glasgow home. Dugald Clerk enjoyed a long and distinguished career in engine research, while it was the English engineer, Joseph Day, who in 1889 took Clerk's two-stroke a stage further. Day developed the two-stroke engine so that it became oil-fuelled and could therefore be made smaller, and this, of course, was where the two-stroke would find its niche; compact and mechanically simple it proved ideal for powering equipment such as lawnmowers, chainsaws and outboard motors – indeed, almost anything that required a source of high power while being easily transportable, and the two-stroke became increasingly popular with men looking for any excuse to get out of the house.

The two-stroke engine's erstwhile wide use in motorcycles was phased out in the 1980s when, because of its low fuel efficiency and high pollution level, as well as its ear-splitting volume, it was replaced by the four-stroke and other engines which, if not exactly green were vaguely turquoise in comparison. While its inefficiency made it generally unsuitable for the car industry, two-stroke cars have been produced with the most famous perhaps being the East German Trabant which was the car of no choice for the population of the former German Democratic Republic from 1958 until 1989 – even though it remained consistently slower than most of East Germany's female athletes.

Healthy Scots

Residents of the County of Lanarkshire were outraged and the rest of the country shocked when a survey in April 2010 claimed that more than twenty percent of people in Coatbridge were obese, making Coatbridge the fattest town in the UK – and probably somewhere in the world top ten. This caused outrage in Lanarkshire because of the implication that Coatbridge's take-away establishments must be offering larger portions than anywhere in the region, which as far as the good people of Motherwell, Hamilton and so on were concerned begged the question why were their portions smaller?

Through the centuries Scottish physicians, surgeons and nurses have led the field in medical innovations which have transformed and indeed saved millions of lives the world over. When one considers the apparent contradiction between this noble history to further advance health and medicine and the results of modern Scottish living the word that springs to mind is irony – except in Coatbridge where irony is of course someone who drinks too much Irn-Bru.

CAN I HAVE MY FIVE-A-DAY WITH CHIPS PLEASE?

61 JAMES LIND (1716–94)

The Scots in Scotland have become infamous for having a bad diet. For all the wonderful, fresh produce grown and reared on its farmland and crofts or caught in the sea, the reputation of Scottish cuisine as a heart attack waiting to happen is well established. Yet it was a Scot who in 1747 made one of the greatest discoveries in the history of nutrition and saved thousands of lives as a result.

James Lind was born in Edinburgh, studied medicine and served in the Royal Navy as a surgeon. For all mariners at sea, scurvy was a killer. The longer the voyage and the farther away they were from fresh fruit and vegetables, the danger of succumbing to scurvy were increased, and once the tell-tale signs of fatigue, spots and bleeding gums began to show themselves no amount of rum, sodomy or the lash could reverse the process. The Royal Navy, being one of the largest in the world, suffered more than most but nobody knew how to prevent scurvy. Different remedies had been suggested over the centuries but none had proved definitively successful. In 1747, Lind set about running the world's first recorded clinical trial. Sailors with scurvy were divided into groups and each was given a different remedy ranging from cider, to vinegar, to seawater, to fruit. The only group that showed definite improvement was the one that had been given oranges and lemons, and in 1753 Lind reported his findings in *A Treatise of the Scurvy* and recommended that citrus fruit should become a regular part of the diet at sea.

Lind spent the rest of his life campaigning for improved conditions at sea including better hygiene and cleanliness on board, which reduced the incidence of typhoid. However, it would not be until 1795, a year after Lind's death, for a sceptical and parsimonious Royal Navy, reluctant to spend additional money on supplies, took note – and even then provided lime rather than lemon juice as it was on a special, half-price offer. This led to the Brits being nicknamed Limeys by their American counterparts, which was meant to be a term of derision, but as the alternative to drinking lime juice was often swimming with the fishes, being called a little fruity from time to time was not too much of a hardship.

The consumption of fruit juice became accepted practice in the 19th century, first as a cure and then as a preventive measure, and dramatically reduced the incidence of scurvy although, still, nobody knew why this was so. It was not until 1912 that the concept of vitamins was proposed as a constituent of the nutrients and acids essential to human health and more than another twenty years until the ascorbic acid present in foods, including citrus fruit, was isolated in 1933 and given the name vitamin C, leading to an increase in the sales of oranges, limes and tangerines and a greater demand for having one's lemon squeezed – purely for health reasons of course.

In 1797, two years after the Royal Navy had adopted Lind's recommendations, a consignment of Seville oranges arrived in Dundee and was bought by a local merchant, James Keiller. The Spanish boat had sold them at a bargain price and Keiller only later discovered that the oranges were too bitter to eat. He took the highly unusual and extremely dangerous step of informing his wife about this financial disaster whereupon Janet Keiller took charge and made the fruit into a jelly, throwing in strips of the peel for good measure. Janet's extremely grateful husband realised that such a recipe could be sold as a less sweet alternative to other jams and preserves and to show his appreciation bought his wife several more consignments of oranges and a new apron.

And while on the subject of the nation's love for the sweeter things in life, in 1883 Greenock-born businessman and shipping company owner Abram Lyle saw an opportunity for the treacly substance produced when refining sugar at his factory in East London and in 1885 launched Golden Syrup which remains as popular today as it did 125 years ago, even if its famous biblical logo of bees swarming around the body of a dead lion is not particularly conducive to delicious home baking.

THE DRUGS DO WORK

62 JAMES YOUNG SIMPSON (1811–70)

By the late-18th century Scotland and in particular Edinburgh University was at the forefront of excellence in medicine and anatomy, having placed the disciplines on a fully scientific footing and, Scots being Scots, once they had gained their degrees many graduates departed to make their name and their fortune elsewhere. The world's two most prominent anatomists were brothers, John and William Hunter, from near East Kilbride; both trained in Scotland and both made their names in

London – John for his research in dentistry and venereal disease, although thankfully not at the same time, while William, who had specialised in obstetrics, became the personal physician of Queen Charlotte, wife of George III.

John Hunter was ever keen to push back the boundaries of knowledge by experimentation as well as observation and his familiar adage was, 'Don't think, try', although when he accidentally injected himself with gonorrhoea he probably wished he had done less of the latter and more of the former. Nevertheless, he was the teacher and the good friend of English scientist Edward Jenner who had been greatly influenced by Hunter's work and methods and in the 1790s Jenner discovered a vaccination which was successful against smallpox, one of the greatest breakthroughs in medical history.

The Hunter boys were part of the trend whereby having a Scottish doctor at your beck and call was the ultimate accessory for heads of state: William was the personal physician of Queen Charlotte, while his brother John held the same position for King George III; James Craik from Kirkcudbright, who emigrated to America, was the long-term personal doctor to and close friend of the first US President George Washington and, most curiously, James Wylie from Fife who studied medicine at Edinburgh without graduating ended up at the Royal

Imperial Court of Russia where he became personal physician to one of Russia's most notable Tsars, Alexander I, from 1801–25.

By the 19th century Scottish surgeons had become so proficient that they could successfully remove a limb in the time it took to boil a kettle, although the patient would usually require something stronger than a cup of tea and a biscuit afterwards. James Syme from Edinburgh was a prominent surgeon with a considerable number of claims to fame including: successfully removing a four-pound (2-kg) tumour from a man's jaw without an anaesthetic; discovering that dissolving rubber in naphtha led to waterproofing, while leaving Charles Macintosh [5] with the patent, the honour and the financial rewards for inventing the Mackintosh raincoat and, demonstrating slightly better foresight, hired Englishman Joseph Lister as his assistant in 1854.

Lister went on to become one of the most important figures in the history of preventative medicine with his pioneering work in the 1860s as professor of surgery at Glasgow University and Glasgow Royal Infirmary where he introduced antiseptic techniques in the operating theatre alongside a range of measures aimed at keeping hospital wards as clean as possible. Syme supported and encouraged Lister throughout and the bond between the two men strengthened when Lister fell in love with

Syme's daughter, Agnes, and asked for her hand in marriage, to which Syme said 'Yes' – as long as Lister ensured that he washed his hands first.

During his illustrious career James Syme clashed with almost everybody and two among his profession who learned to steer clear of his scalpel were the surgeon Robert Liston from Linlithgow, and the physician James Young Simpson who was born in Bathgate, West Lothian. When in 1846 American dentist William T. G. Morton, who boasted Scottish ancestry, first used ether as an anaesthetic, doctors in Scotland were obviously keen to try it, too. Liston – Syme's rival as the fastest knife in the East – was the first surgeon in Europe to use ether successfully on a patient and Simpson also used ether early in 1847.

But James Young Simpson, obstetrician and for thirty years Professor of Midwifery at Edinburgh University, elected to turn his attention to chloroform also recently discovered to have anaesthetic properties. Following John Hunter's tradition of experimentation and observation, he would host dinner parties at which, once the main course was over, he would administer varying doses of chloroform to himself and a few brave guests – who would have preferred some opium but what the hell – and note the effects. Initially, finding out how much to give was a matter of trial and error and on more than one occasion Simpson's housekeeper and cook would find the doctor and his friends unconscious, before to the cook's great relief realising that the bread-and-butter pudding was completely untouched.

But Simpson continued to hone his experiments and in November 1847, an anaesthetic was used to relieve pain during childbirth – a world first –with both mother and newly born daughter safely surviving the procedure, although the poor child could have done without being named Anaesthesia in honour of this new medical advance.

There was much criticism of Simpson's use of anaesthesia with some, such as Syme, thinking it too dangerous while others questioned Simpson's moral right to interfere with natural pain of childbirth, it was the Will of God after all. Interestingly, almost all of this criticism came from men and as far as women were concerned the more gas the better. Official recognition for the use of chloroform came from the highest power in the land when Queen Victoria, who had shown that she was not too posh to push having born seven children to date, insisted on anaesthetic pain relief for the births of her last two children, Prince Leopold and Princess Beatrice in 1853 and 1857 respectively – and if it was good enough for Queen Victoria it was good enough for the millions of other women.

The use of chloroform as an anaesthetic would eventually fall out of favour in the 20th century, but the use of anaesthesia in childbirth was here to stay, no matter what nature, God or Will had to say about it. Simpson would go on to have the words *Victo Dolore* – pain conquered – inscribed on his coat of arms, which was, of course, not strictly true, but considering what childbirth had been like before anaesthetics most people forgave Simpson his exaggeration.

BIKE? I'LL GIVE YOU BIKE

63 KIRKPATRICK MACMILLAN (1812–78)

The Olympics is a strange institution. Every four years for three weeks the nation's attention is fixed on the greatest sportsmen and sportswomen in the world, and enthralled by the determination, agility, power and speed of smiley Sue Barker, perky Hazel Irvine and the rest of the BBC team's efforts to describe and explain the skills, intricacies and subtleties of more than twenty-five sports that three weeks earlier nobody, including the commentators themselves, knew anything about. Glory and gold go to the victors; despair and desolation are inflicted on the vanquished. National heroes are made, Union Jacks are flown and once it is all over and David Beckham has concluded the closing ceremony by extinguishing the Olympic flame with a trademark free-kick, the games are immediately forgotten as life returns to normal and thoughts return to who is going to captain your fantasy premier league football team.

Yet Olympic champions are not entirely forgotten, and every four years are brought back out of retirement to relive those glorious moments and provide expert analysis on the Greco-Roman wrestling finals – a phenomenon not unlike the one which takes place when voting for Scottish MPs in general elections, although in this case the recognition factor is often restricted to the short period spent in the polling booth, when only by scanning the names on the ballot paper are you finally reminded who your MP has been for the past five years.

Scotland's undisputed greatest Olympian is Edinburgh-born cyclist Chris Hoy, who with four golds and one silver medal from three Olympics is also the greatest-equal Olympic cyclist of all time; his three gold medals at the 2008 Beijing games put him on a par with the entire Brazilian Olympic team leading to Scottish calls for some form of play-off at Portobello between Hoy and the Brazilian beach-volleyball team.

It is of course entirely appropriate that Scotland's greatest Olympian happens to be a cyclist as it was in Scotland that the modern-day bicycle was born. The first 'bicycle' was invented by a German, Karl von Drais, around 1817, was called a draisine and had two wheels and a frame that you straddled and walked along with – because it had no pedals. Momentum was gained and if you were lucky you would achieve a gliding-walking action, with changes of direction enabled by the front wheel that could be steered. These draisines were meant for perambulating around your garden – if you had a garden – were made out of wood and briefly became all the rage throughout Europe in a craze not dissimilar to the inexplicable success of the space hopper 150 years later.

In Britain, the draisines were nicknamed hobbyhorses, but they quickly became nothing more than an unfashionable toy. Then, in the 1860s various French inventors came up with and patented a two-wheeled machine with pedals and a crank to give power and momentum. This, with a wooden frame and iron bands on the wheels, began to be mass-produced in 1868. The new bicycle was called the velocipede, meaning fast foot, but became known as the boneshaker with regard to the experience of riding on its un-sprung iron frame rather than the experience of when you fell off. Precisely who was the first to produce the fully functioning, working model has never been clearly ascertained, but we were at least safe in the knowledge that the inventor was French. Or so we thought . . .

There were several more incarnations in the ensuing

decades – including the magnificent penny-farthing with its five-foot-diameter front wheel – before, in the 1890s, with smaller, safer designs and the addition of the pneumatic tyre invented or re-invented by John Boyd Dunlop [58], bicycles finally took over the world – and it was at that point that Scots said, 'Hold your hobby-horses'. It was not, proclaimed the Scots, the French who invented one of the most important means of land transportation, it was, in fact, a blacksmith from Dumfries and Galloway.

Kirkpatrick Macmillan had, in his youth, seen one of the hobbyhorses and determined that he could come up with something that did not involve all that walking; in 1839 he began to build his own version. The machine he came up with was wooden with two iron-rimmed wheels, the front wheel could be steered and there were pedals connected by rods to a crank on the rear wheel: he had built the first pedal-driven bicycle in the world. Macmillan was delighted with his invention and although the bicycle was exceedingly heavy and the country roads were not great, he would regularly ride his bicycle the fourteen miles (22.5 km) from his smithy in the village of Keir to Dumfries – and with the wind behind him and no cows on the road he could complete the journey in under an hour.

In 1842 he rode his bicycle sixty-eight miles (109 km) from Keir to Glasgow, a journey that took Macmillan two days to complete. But his considerable achievement was marred somewhat when he knocked over a small girl who ran across his path in the Gorbals and he was fined five Scots shillings for 'furious driving' and general negligence while in charge of a large wooden moving thing; but at least when he returned to his wooden bicycle he was relieved to see that it was still where he left it – although there was clear evidence of singeing.

Macmillan was an unassuming and modest man and after his adventure in Glasgow he was happy to spend the rest of his days in his beloved Dumfriesshire. He had no interest in patenting or publicising his bicycle, and although he did occasionally make bikes for other people he had no interest in becoming involved in commercial bicycle production.

There is a further twist in the tale of the bicycle. For a long time, it was thought that Gavin Dalzell from Lanarkshire, who began building bicycles in the 1840s, was the original inventor. Even now, due to the lack of contemporary recorded documentation, there is considerable controversy over Macmillan's biography and the claims of origination made on Macmillan's behalf. Not that this would have bothered Macmillan in the slightest; he was more than happy cycling back and forth to Dumfries, and this most modest of men would no doubt be slightly embarrassed by the commemorative plaque at what used to be the Macmillan smithy inscribed, 'He builded better than he knew' – and particularly embarrassed by the past tense form of 'build'.

Despite a century of competition from the motor car and the lack of the bicycle helmet's aestheticism, more than one hundred million new bicycles are produced every year – that's more than five times the number of new cars – and there are believed to be more than a billion bicycles in the world, and more than a billion bicycle owners. And with the bicycle providing a healthier, cheaper and more environmentally friendly alternative to the automobile it is almost guaranteed to play as integral a part in the future of the transport as it has in the past and does in the present.

Scotland is as you can imagine incredibly proud of the bicycle: it was invented by a Scot, and the fastest cyclist in the world is a Scot and just over half the households in Scotland own bicycles which one day –who knows – they might even get around to taking out of the garage.

ENVIRONMENTALISM: WHAT BEARS DO IN THE WOODS SHOULD STAY IN THE WOODS

64 JOHN MUIR (1838–1914)

East Lothian is an often-neglected region of Scotland. When visitors arrive in Edinburgh and inquire about a day trip out of the capital it is usually a fourteen-hour bus trip to Loch Ness that they have in mind, or if they are being particularly brave head west to Glasgow. Precious few are aware of the historic towns, rolling countryside and stunning beaches on Edinburgh's doorstep. For those who do venture into East Lothian, then one of the jewels of the countryside is the John Muir Country Park which includes woodland, grassland, coastline, several species of duck and is blessed by being situated near the town of Dunbar, which due to receiving more hours of sunshine than any other town in Scotland revels in the nickname, Sunny Dunny.

The country park is named after Dunbar-born naturalist and conservationist, John Muir, and impressive although its 2.7 square miles are even the most dedicated of the park's devotees have to admit that East Lothian's finest is slightly overshadowed by the 907 square miles of the John Muir Wilderness in the Sierra Nevada, California. This vast and varied landscape supports an abundance of wildlife from golden trout to golden eagles and the Jeffrey pine – which was named after Perthshire-born botanist, John Jeffrey.

Muir is often referred to as the Father of America's National Parks. He emigrated with his family in 1849 when he was eleven years old and settled in Wisconsin and it was there that Muir studied geology and botany at university. He was fascinated by natural history and made it his life's work, travelling all over the US and Canada before settling in California. In this regard Muir's inclination westward has typified that of so many of his fellow countrymen in America at the time, and California now has more Scottish descendants than any other US state.

The Sierra Nevada and Yosemite became Muir's playground and right from the start he campaigned tirelessly for the area to be protected from ever-increasing agricultural development. Eventually, in 1890, Yosemite Valley was declared a national park, the second in America after Yellowstone, and with Muir agreeing to perform the role of faithful Old Geyser. Not satisfied with the preservation of Yosemite, two years later Muir co-founded the Sierra Club, which is America's oldest and longest-running environmental organisation to date, and remained its president until his death.

John Muir ranks as one of the key figures in the history of world ecology and environmentalism with his fervent belief that it is incumbent upon mankind to protect nature's wilderness for the benefit of future generations. In addition to the John Muir Wilderness in his beloved Sierra Nevada, Muir's name appends the region's famous, 200-mile trail – one of the most spectacular wilderness walks in the world – while close to the trail stands Mount Muir, which rises to more than 14,000 feet (4,267 m).

An indication of Muir's celebrity and importance was that in 1903 when the new American President Theodore Roosevelt, who was renowned for his love of the great outdoors, went to see Yosemite for himself, John Muir was his guide and the two men, out in the open air and with common interests in the land, talked into the early hours. It was a night that neither man would ever forget, not, I hasten to add that it was anything like *Brokeback Mountain*, as *Brokeback Mountain* is of course in Wyoming and not California.

GIVE ME FEVER

65 PATRICK MANSON (1844–1922)

Did you hear the one about the Irish doctor, the Scottish doctor and the French doctor who all thought they had invented the hypodermic syringe? The Irish doctor, Francis Rynd, thought he was the inventor as in 1844 he was the first to come up with a needle – it was wooden – that could enter a vein; the Scottish doctor, Alexander Wood, thought he was the inventor as in 1853 he was the first to administer drugs using a metal needle and the French doctor, Charles Praza, thought he had wandered into someone else's joke.

In fact all three doctors had a claim on the invention of the hypodermic needle as it is impossible to say for sure who was the first; however many people opt for Wood out of sympathy on the basis that his wife, who volunteered to be one of her husband's guinea pigs, ended up with a serious morphine habit and died of a self-administered overdose. This was obviously deeply upsetting and traumatic for Dr Wood, but on the positive side it did show that his invention worked.

During the second half of the 19th century Scottish-trained doctors remained in high demand around the world and one who travelled far from home was Patrick Manson, from the Aberdeenshire village of Oldmeldrum. After graduating from Aberdeen University Manson found himself a job in China where from 1867 he worked for twenty-three years, first in Taiwan (then Formosa) where he was employed as a medical officer and then as a doctor in the city of Xiamen on the mainland.

In 1883 Manson moved to Hong Kong, shipped out some cows from Scotland to improve the local dairy herd and, three years later, set up a company called Dairy Farm – now owned by Jardine Matheson [William Jardine, 24] and is today one of the largest retail companies in Asia. The following year, 1887, he was made Dean of College at the newly founded Hong Kong

College of Medicine for Chinese. One of Manson's first two students was Sun Yat-sen, a future revolutionary, first president of the Chinese Republic, Father of the Nation and thanks to his training at the college a fully qualified doctor with the prerequisite illegible handwriting.

While his achievements in Hong Kong were considerable, Manson is most readily remembered today for his pioneering and extensive work in tropical medicine. Of all the diseases prevalent in China – which included both cholera and typhoid – malaria was the biggest killer, in fact, malaria was the biggest killer in the world and Manson devoted his life's work to determining its cause. His first breakthrough came in 1877 when he was treating patients who had elephantiasis – a condition contracted from a parasitic worm that causes massive, ugly swelling of, usually, an arm or leg. Manson noticed that the worms he found in his patients' blood samples were far more prevalent at night and became convinced that there had to be some connection between the worms and mosquitoes who were also more prevalent at night.

Manson persuaded his gardener, who was already infected with elephantiasis to allow mosquitoes to feed on him at night so that the mosquitoes could be caught and dissected and discovered that the insect did indeed carry the worms. Manson was delighted as he had finally shown that there was indeed a link between elephantiasis and mosquitoes and in recognition of his gardener's bravery gave him a couple of hours off in the morning.

After making one link between tropical diseases and the mosquito Manson painstakingly continued his research, finally returning to Britain in 1889, and in 1894 he published a comprehensive study detailing his work so far and developing his theory that there was also a link between mosquito and malaria. At this point Manson did not have the evidence he required to prove such a link, and many scientists were critical of his theory. Undaunted, Manson continued his research, but it was another physician Ronald Ross who in India in 1898 and following much of Manson's hypothesis would finally find the definitive link between the two. To prove this medical breakthrough beyond doubt mosquitoes were sent to London and Manson's son, a medical student, was subjected to the little blighters and within two weeks had contracted malaria – one suspects that Manson was particularly generous with his son's allowance that year.

Confirming the connection between malaria and the mosquito has not eradicated the disease, malaria is still contracted by over 200 million per year, but it did mean that scientists and doctors could begin to treat the symptoms and work on preventative measures such as anti-malarial drugs – and, not least, promoting the use of mosquito nets – which have saved many, many millions of lives, seen the eradication of malaria from North America, southern Europe and many other parts of the world and seen actual fatalities fall to less than a million a year.

For his work on this killer disease, Ronald Ross gained the Nobel Prize for Medicine – although this was controversial as other physicians and scientists had played important roles in the discovery and all had been working during the same time. Manson, whose contribution to the findings had been far from insignificant, founded the London School of Tropical Medicine in 1899 and after a lifetime spent studying little biting insects will ever be known as the Father of Tropical Medicine. Another Scot who played a critical role in the control of disease was Glasgow-born William Boog Leishman who, in 1901, discovered the parasite that caused kala azar or 'black fever', named Leishmaniasis in his honour. He also, after working with Almroth Edward Wright who developed the typhoid vaccine, was responsible for the fact that soldiers in the British Army were inoculated against typhoid in 1914 at the outset of World War I. Despite the appalling conditions that the British army faced in the trenches, where the lack of adequate sanitation and hygiene would normally have been ripe for the spread of the disease, only 1,100 British soldiers would die of typhoid in four years. So, as much as you could blame the disastrous events of World War I on the politicians and generals, or even on the not so good times, you could not blame it on the Boogie.

DIABETIC DOGS – ONE OF DAVID BOWIE'S LESSER-KNOWN ALBUMS

66 JOHN JAMES RICHARD MACLEOD (1876–1935)

It is estimated that just short of three percent of the population of the world, or 170 million, people suffers from diabetes and that figure is set to increase dramatically in the next couple of decades, not primarily because of sedentary lifestyles and unhealthy diets (although that does not help), but because the world's population is growing older. However, for all the millions that suffer from diabetes and the complications that the disease brings, the great majority of diabetics find that with straightforward drug treatment and careful management of their diet they can live a normal life – and all thanks to a medical breakthrough in Canada in

1921 that was in no small part down to a doctor and professor from Cluny near Dunkeld.

J. J. R. Macleod studied medicine at Aberdeen University, worked in London, and in 1903 became Professor of Physiology at what is now the Case Western Reserve University in Cleveland, Ohio, before in 1918 accepting a similar position at the University of Toronto in Canada. Macleod's speciality was carbohydrate metabolism and how the body breaks down food in the process of digestion and converts it into energy, and through his teaching, lectures and numerous books he became an internationally respected figure on the subject. It was this renown that led in 1921 to Canadian scientist Frederick Banting approaching Macleod to ask for backing and university laboratory space in which to conduct an experiment which, if his theory was correct, might lead to a cure for diabetes.

In the early 20th century irrespective of what form of diabetes you had anyone unfortunate enough to be afflicted by it could expect to suffer from symptoms including blindness, kidney failure and a very short life expectancy. There was no known cure and the only treatment on offer was a starvation diet, which might hold down the patient's glucose level but was likely to finish them off through malnutrition. It had been established that a build up of glucose in the blood caused the symptoms of diabetes and that blood glucose levels were regulated by a hormone in the pancreas; if that hormone was absent (or not getting into the bloodstream) then the blood glucose level would rise. The hormone, a protein, had been named insulin, and it was known to be stored and released when needed by cells called the islets of Langerhans, which although they sound like a German holiday resort are actually found in the pancreas, but no one had been able to be isolate the hormone or even throw a German beach towel over it.

Banting wanted to conduct an experiment on dogs. He would remove the dogs' pancreas and before long the dogs would develop diabetes and then he would inject the dogs with a pancreas extract to see if the diabetic symptoms of the dogs would improve. Macleod was initially sceptical about Banting's plan, but he was going off on holiday for the summer so, what was the harm in letting Banting have laboratory space for a few months? He loaned Banting an assistant, Charles Best, plus ten dogs on the proviso that the laboratory would be properly fumigated before his return, and if he found even one stray dog hair then there would be trouble.

The experiment did not begin well as the dogs kept dying before the pancreatic extract could be administered or for the dogs who managed to initially survive the removal of their pancreas, the first extracts administered ended up poisoning them and soon it was not a case of Macleod returning to a stray dog hair, but a whole laboratory of dead stray dogs. Throughout that long, hot summer of 1921 there were a series of unexplained dog-napping incidents in Toronto and many upset children worried about the prospect of their beloved family pet being the next to disappear, which mysteriously always appeared to be somewhere near the university.

When Macleod returned to Toronto, he discovered that as well every dog in the city appearing to be on a lead, Banting and Best had made a breakthrough and a dog named Alpha had been kept alive. Macleod was impressed by the men's work, but found errors in their conclusions. He took control of the team, overseeing the work, and ensured that there were more experiments, more finance and more detailed and thorough analysis. And he brought in biochemist James Collip at the end of 1921 expressly to come up with a way of purifying insulin extracted from cattle and pigs so that it could be used safely in humans.

In January 1922 the first human trial took place on a fourteen-year-old boy critically ill with diabetes, and after two injections of Collip's extract the boy showed significant improvement. Further trials also conclusively proved that insulin while not curing diabetes could significantly improve patients' symptoms. In 1923 insulin was produced commercially for the first time and diabetics began to benefit from the trials on those unlucky dogs, and especially Alpha – who had endured so much, but would sadly never be a Romeo again.

The enormity of the team's achievement was recognised immediately throughout the world and the 1923 Nobel Prize for Medicine was awarded jointly to Banting and Macleod. This caused considerable friction with Banting announcing that he would share his prize money with Best, and Macleod saying that he would share his prize money with Collip. Banting and his supporters were dismissive of Macleod's involvement in the discovery of insulin, and Macleod returned to Aberdeen University in 1928. J. J. R. Macleod's name is not the one that tends to be remembered for insulin, but it was Macleod who was the team leader, it was he who ensured that all the conclusions were comprehensively backed up and it was he who brought in Collip.

The final disharmony made a sad epilogue for the four men who as a team had made one of the most important medical breakthroughs of the century, and it was a far cry from their altruistic decision to sell the original patent to the University of Toronto for only one dollar, so that diabetics everywhere could benefit from their discovery.

That the discovery of insulin was important goes without saying, and another co-Scottish discovery in 1957 may prove at least as momentous. Virologist and biologist Alick Isaacs was born in Glasgow and became an expert on influenza. In 1950 he became director of the World Influenza Centre where he studied a huge variety of strains of flu and the various ways that the body tried to combat the virus. In 1957, Isaacs together with Swiss scientist Jean Lindenmann Isaacs, discovered that, under attack from a flu virus, cells in a hen's egg produce a protein that tries to stop, or interfere, with the virus. Isaacs and Lindenmann published their results to international acclaim and gave this protein the name, interferon.

It was hoped initially that interferon would be a wonder drug, capable of curing all manner of illnesses, but progress was slow and it would not be until the 1980s, long after Isaac's early death in 1967, that interferon began to be prescribed in medical treatment. Today, interferon is used to help treat multiple sclerosis, hepatitis and some forms of cancer and it is hoped that with further research future generations will see the full health benefits of yet another groundbreaking discovery from a citizen of the city that gave us chips and cheese and curry sauce.

HOW CLEAN IS YOUR LAB?

67 ALEXANDER FLEMING (1881–1955)

Scottish men are not noted for their contributions toward chores around the house. When Sarah Brown once publicly commented that her husband was 'messy'this gentlest of admonishments made the Prime Minister, as he was then, seem for once like a regular sort of guy, only for Gordon to slightly spoil it by replying that, while he did take responsibility for being slightly untidy, he had in real terms done more washing up in the last financial year and thirty percent more vacuuming than when the Conservatives were in power. What Gordon Brown [30] might have pointed to instead, was the example of the Scot who arguably made more of an impact internationally in the 20th century than any other – and the fact that the discovery that he is known for came about primarily because he did not wash his dishes.

It would take Alexander Fleming many years to become an overnight success and the virtue of patience is a recurring theme of his life and career, which may have had something to do with his father being aged sixty-five when baby Alexander was born on an East Ayrshire farm in 1881. Fleming went to school in Kilmarnock and at the age of fourteen moved to London where he found work as a shipping clerk. It would not be until he was twenty that he won a scholarship to study medicine at St Mary's Hospital Medical School in London from which he qualified in 1906 and, having proven himself an outstanding student, was given a full-time job at St Mary's specialising in bacteriology. During World War I, his experience serving on the Front with the Army Medical Corps and seeing hundreds of deaths due to infected wounds inspired Fleming to redouble his efforts in the fight against infection.

Fleming became a brilliant bacteriologist and in 1922 discovered enzyme lysozyme, the protein that is found in tears. Six years later, in the summer of 1928 Fleming had been studying staphylococcus bacteria and had kept all his culture cells in Petri dishes. With the holidays coming up, all Fleming had to do before he went away was make sure he had tidied up behind him. However, doing the dishes was not one of his fortes and his idea of tidying up was the traditional Scottish male method of piling up all the dishes in a corner. In his haste, he failed to notice that the lid on one of the dishes had not been carefully secured, and when he returned from his hols in September 1928 he noticed a strange blue-green mould had grown on the culture in the dish. He nearly threw the corrupted dish into the bin, but noticed something was different and on further investigation realised that the mould had killed of all the bacteria in the dish it had come in contact with. In one of the greatest understatements in history Fleming said, 'One sometimes finds what one is not looking for.'

Alexander Fleming called his mould penicillin from the Latin *penicillium* and wrote a paper in 1929 about its antibacterial properties, to which there was generally complete indifference. He continued to labour on penicillin for several more years but was unable to discover how to produce enough of the antibiotic to make it commercial or, more fundamentally, how to get it to remain in the human bloodstream long enough to have a significant effect. Eventually Fleming had to abandon penicillin and move on to other research, although he did have the consolation of becoming Professor of Bacteriology at the University of London in 1938, even if this did not necessarily make him any tidier.

In 1939 with the outbreak of World War II it became imperative to try to make antibiotics viable in order to avoid the high death toll from infections that had marked World War I. The year before war broke out, a team of Oxford University scientists led by Howard Florey and Ernst Chaim had returned to Fleming's work on penicillin and, in 1941, they made the crucial breakthrough that enabled this natural antibiotic to be improved, tested, synthesised and produced in a form effective in the treatment of infections in humans. Penicillin as a mass-produced drug was first made available to the US army in 1944, in time for the Allied invasion of Europe, and thousands of lives were saved as a result.

Once the war was over the world's first and best-known antibiotic was made available internationally for civilian use and, ever since, has been instrumental in saving countless lives. Penicillin also resulted in the development of what would become the immensely lucrative pharmaceutical industry.

Alexander Fleming was never invited to work with Florey and Chaim and the rest of their Oxford team on penicillin and was always modest about his role compared to theirs; but at the same time it was never forgotten that Fleming was the original, six million dollar antibiotic man who discovered penicillin in the first place, and it is he who has been most closely associated with the drug. In 1945, the Nobel Prize for Medicine was awarded to all three men, Florey, Chaim and Fleming. It had been seventeen long years since September 1928 and the day that Fleming had returned from his holiday to be confronted by his dirty dishes. Which only made the perennially patient if hygienically liberal Fleming ponder what medical miracles he might have found in his laboratory if rather than going on holiday for a fortnight he had gone for a whole month.

THAT SWIMWEAR IS SO MANLY

68 ALEXANDER MACRAE (1888–1938)

There is considerable speculation about how the name Harris Tweed came into being, considering that the Isle of Harris in the Outer Hebrides and the River Tweed in the Borders are geographically so very far apart. Both Harris and the Borders have for centuries been known for their textiles and it is possible that the names were simply put together to describe a generically Scottish cloth; another theory suggests that 'tweed' was mistakenly used for 'tweel', a Scots word for patterned cloth. Whatever the case, from the middle of the 19th century Harris tweed, as in hand-woven from local, Hebridean wool, began to be made into jackets, suits and hats and marketed to the rich and famous and those who aspired to be rich and famous; the Royal Family, the landed gentry, golfers and Sherlock Holmes all wore Harris tweed. But by the mid-20th century tweed had gone out of fashion and the term 'tweedy' had such a negative association that even marriage to a Chelsea footballer was seen as being preferable. However, the fabric has seen a recent renaissance as a timeless symbol of quality and was worn with pride by Tom Hanks in the *Da Vinci Code* films, Matt Smith in *Doctor Who* and by the instigator of its revival on screen, Clint Eastwood, in the *Dirty Harry* movies – but there is no truth in the rumour that he once visited a Harris croft and said, 'Come on sheep, make my jacket'.

Most people however, associate Scotland with tartan and the kilt, and no international representation of the nation is complete without Scotsmen baring their legs, and sometimes considerably more, while attempting to foster diplomatic relations or at the very least obtain a free round of drinks. Of course this nationwide appropriation of Highland dress is a phenomenon dating back only as far as the 19th century and for which the romantic novels of, among others, Walter Scott [35] were largely responsible. Irrespective of the historical validity of Lowland Scots taking ownership of a Highland culture that for centuries they attempted to marginalise, the wearing of the kilt does show a propensity of Scottish men in general to be as unencumbered as possible as far as their apparel is concerned and, therefore, it should not be too surprising to learn that the man behind the world's most famous swimwear company was born in Scotland – near the Kyle of Lochalsh in the West Highlands.

Businessman Alexander MacRae immigrated to Sydney in 1910 where, four years later, he set up MacRae Hosiery specialising in the manufacture of socks and underwear. His company prospered and with a more liberal Australia in the 1920s heralding the beginning of a beach culture and mixed bathing, MacRae saw the potential in producing swimwear for young Sydneysiders keen to show off their doggy paddle. In 1928 he developed a new, figure-hugging, non-wool swimsuit for men and women; the Racerback was designed with athleticism in mind and had an open, strapped back for ease of movement. To coincide with this development MacRae wanted a new name for his company and organised a staff competition to find one, the winning slogan was, Speed On in Your Speedos', and an international market leader was born.

In the 1932 Los Angeles Olympics, when Australian Claire Dennis, wearing a Racerback, won the 200 metres breaststroke traditionalists wanted her disqualified on the grounds that her costume showed 'too much shoulder' – which considering how big swimmers shoulders tend to be must have been quite a shock back then. Nevertheless, Speedos became the swimsuits of choice of the Australian swimming team and in the 1950s began to be exported. With the launch of its classic, swimming briefs in 1961, Speedo became the world's biggest brand name in swimwear.

Alexander MacRae died in 1938 and thus missed out on global domination, but he did live long enough to see his designs revolutionise both sport and leisure;

although when it comes to swimming in the Highlands, no matter how much extra movement and speed your costume may afford, it is still bloody freezing.

WHO'S THE DADDY?

69 IAN DONALD (1910–87)

Glasgow has for many years been burdened with the unwanted tag the 'sick man of Europe', on account of Glaswegians' historically poor public health record and low life expectancy in relation to the rest of the UK and continental Europe. But disappointingly it seems that no matter how much time, effort and money is invested, Glasgow seems no closer to gaining the world rather than the European title

At the same time, Glasgow has since the 18th century, when both William Cullen [71] and Joseph Black were professors of medicine at the city's university, played its part at the cutting edge of medical innovation, which makes you wonder if, without all those sick Glaswegians to practise on, might otherwise not have been the case.

It was while he was professor of surgery at Glasgow Royal Infirmary that Englishman Joseph Lister in 1867 published his work on the use of antiseptics. Among the many doctors who took up Lister's principals of antisepsis was surgeon William MacEwen from Rothesay on the Isle of Bute. MacEwen graduated from Glasgow University and, following in Lister's footsteps, became professor of medicine at Glasgow Royal. He was a pioneer in both neurosurgery and bone graft techniques and between 1876-80 his surgical successes included the removal of a brain tumour, a brain abscess and a viable bone graft and in 1895 he became the first to successfully remove a patient's lung.

1895 was also the year that German physicist Wilhelm Rontgen invented X-rays and Glasgow Royal Infirmary was once more at the forefront when Glasgow-born doctor John MacIntyre set up the world's first radiology unit in 1896. In the same year MacIntyre also produced the first moving film of an x-ray, when he filmed the movement of a frog's leg –much to the surprise of the unsuspecting frog, who had only come in because he had a touch of the flu. The city made its medical name again in the 1970s when two English-born professors of neurosurgery, Graham Teasdale and Bryan Jennett, while working at Glasgow University came up with a scale for defining a patient's state of consciousness and the Glasgow Coma Scale (GCS) remains in use throughout the world to this day.

Any discussion about Glasgow's impact on medicine would be incomplete without recognising the innovation in diagnostic ultrasound in the mid-1950s. The use of ultrasound – high-frequency sound waves beyond human hearing – was pioneered in the 1930s and first used by engineers to detect flaws in metal. The idea that ultrasound might have a medical application was first mooted in the late 1940s and early '50s in the US and Japan and it was established that, with the patient immersed in water ultrasound images could be taken to help diagnose tumours. A Paisley-born doctor – the son and grandson of doctors – with an interest in machines was the man who, alongside recognising the massive diagnostic potential of ultrasound, thought to apply it in the way it has become best known today.

Ian Donald spent his early life in Britain before emigrating with his family to South Africa and then returning to London where he continued the Donald medical tradition by graduating in medicine in 1937. In World War II he served in the RAF and received an award for gallantry; he, also, gained his interest and understanding of sound and radio waves. At the end of the war he began a career in obstetrics and gynaecology before moving to Scotland in 1954 when he was appointed professor of midwifery at Glasgow University. In the same year Donald attended a lecture by an English pioneer of medical ultrasound, John Wild, and realised its potential, but knowing that some of his patients bathed only once a week, wouldn't it be better if he could find an alternative to water?

In 1955 Donald visited the engineering firm Babcock and Wilcox in Renfrew, where, armed with several tumours and using a piece of steak, as the control, he used the company's ultrasonic metal flaw-detector that consisted of a probe immersed in oil rather than water to see what images he could define. Early experiments brought mixed results, and nobody would eat Donald's steak pie afterwards, but in conjunction with Glasgow-born technician Tom Brown and fellow obstetrician John MacVicar, Donald persevered. In 1957 when they saved a patient's life by detecting with ultrasound a previously unnoticed cyst, they knew they were making progress and the following year they produced their first two-dimensional ultrasound scanner. Donald wrote up the team's progress and the medical community began to take diagnostic ultrasound seriously.

Included in the findings Donald published in 1958 was the first ultrasound image of a moving foetus and, despite ultrasound's phenomenally wide diagnostic applications, among the healthy general public it is best known for its use on pregnant women. Ultrasound scanners were produced commercially from 1961 and Donald lived to see ultrasound in worldwide, routine use. For so many proud parents-to-be that scan is the unforgettable first snapshot of new life, and the realisation that they have less than six months to renovate the spare room.

THE HEART PILLS OF SCOTLAND

70 JAMES W. BLACK (1924-2010)

Deoxyribonucleic acid is within the cells of all living organisms. It is an intricate and complex molecule, but as far as most of us are concerned all we need to know is that deoxyribonucleic acid is much more commonly known as DNA. DNA stores the genetic code of everything that lives and the *Jeremy Kyle Show* would not exist without it. The discovery of DNA's double helix structure was made in 1953 as a result of the combined work of the English scientists Francis Crick and Rosalind Franklin, American James Watson and New Zealander Maurice Wilkins, and was one of the greatest scientific breakthroughs in history. But it would not have been possible without the contribution of Glasgow-born biochemist Alexander R. Todd who graduated from Glasgow, Frankfurt and Oxford universities and then embarked on a glittering career in academia at Edinburgh, London, Manchester and Cambridge where he investigated the structures of vitamins and synthesised biochemical compounds. At

Cambridge, after World War II, Todd focused on breaking down the chemical structure of DNA and began to put together its nucleotide components, without which knowledge Crick, Franklin, Watson and Wilkins might never have gone on to describe the double helix. Alexander R. Todd was awarded the Nobel Prize for Chemistry in 1957, by which time he had been away from Glasgow long enough to admit that the R actually stood for Robertus.

The Lanarkshire town of Uddingston is famous for being the home of the Tunnock's Teacake, that iconic Scottish delicacy without which no morning or afternoon cup of tea would be complete. Uddingston is also famous for being the birthplace of James Whyte Black, the pharmacologist whose research and development of beta-blockers has eased the symptoms and prolonged and saved the lives of millions of people afflicted with, primarily, heart disease. Any link between these two claims to fame has however yet to be proven.

Black came from a working class family and was brought up in Fife before graduating in medicine from what is now the Dundee University. Black proceeded to follow a career in academic lecturing and research in Malaysia and Glasgow before joining ICI in 1958. It was at Glasgow University that he began his research into the syndrome known as angina. The cause and symptoms of angina had already been described: lack of blood and oxygen reaching the heart, usually through the narrowing of the coronary arteries, leading to an increase in the heart rate, severe and acute chest pain and breathlessness. And it was known that the increase in heart rate, or pulse, was because the hormone adrenaline triggered by beta-receptors caused the heart to beat rapidly in response to the lack of oxygenated blood. The trouble was, the heart beat so rapidly that it became ineffective in pumping out oxygenated blood – and that lack of oxygen was what caused the chest pain and other acute symptoms.

Black set about finding a way to block the action of the beta-receptors and in the late 1950s came up with a drug which did the trick: propranolol. Propranolol was launched commercially in 1964 as the world's first beta-adrenoreceptor blocking drug, or as it is more commonly referred to, beta-blocker. The drug remains in use today, alongside various other beta-blockers developed in its wake, drugs which have transformed the lives of those suffering from heart disease as well as migraine and some stress-related conditions, and have, also, paved the way for even greater consumption of Tunnock's Teacakes.

Black, then, turned his attention to another ailment all too common in the Scots – the stomach ulcer.

Throughout the 1960s and early '70s, Black led the team which figured out how to prevent the hormone histamine from secreting gastric acid – it is the acid irritating the ulcerated lining of the stomach which causes the pain. First they found the receptors that stimulated the production of gastric acid then they came up with a drug to block them. In 1976 cimetidine came on the market and it was a global winner. As well as relieving the pain caused by stomach ulcers it, also, considerably reduced the need for surgery – with less acid eating away at the stomach lining, the ulceration had a chance to heal.

James Black returned to academia in 1973 where he took on professorships in London at University College and then King's. In 1988 he was honoured for his groundbreaking work in pharmacology, and his devotion to easing the suffering of others, by being awarded the Nobel Prize for Medicine. Black had always been modest about his achievements and was completely shocked to win such a prestigious award and by all the media attention that went with it – although at least more than most he would probably know what to take if he felt a little stressed.

Scientific Scots

One wonders when God ever has time in his (or her) busy schedule for a little reflection what he (or she) makes of his (or her) Scottish Presbyterian flock. On the one hand so devout, so fearful, so serious about their faith and for centuries implacable in their determination that Hell would have to freeze over before anybody would interfere with their right to worship and grow extremely long beards. Yet it was scientific Scottish Presbyterians who, with their stubbornness and eternal refusal to bow to authority, that would challenge the very boundaries of the physical and metaphysical world and who would end up dismantling the very concept of divine omniscience itself – except on the Sabbath of course, when they all went to the Kirk.

Considering the extent of the Scots' contributions to science over religion, Scotland the country can count itself extremely fortunate to have been afflicted over the centuries only by plagues of midges and youths wearing casual sports clothes, which only goes to show that, as supreme beings go God is remarkably benevolent.

ICE COLD IN ALLOA

71 WILLIAM CULLEN (1710–90)

For a country not known for having an especially warm climate and where even immigrants from the Baltic find an Aberdonian winter difficult to endure, you would not expect Scotland to have had any interest in anything to do with refrigeration. Yet while the first commercially produced refrigerators were not introduced until the 1920s, the science behind refrigeration was uncovered at Glasgow University nearly 200 years earlier.

William Cullen was born in Hamilton and became Professor of Chemistry at the Glasgow University and Professor of Medicine at Edinburgh. At Edinburgh he lectured and wrote on human diseases and was the first to coin the term 'neurosis' with reference to psychological disorder. Cullen was one of the outstanding scientific lecturers of his time and in Glasgow, in 1748, he demonstrated the first recorded example of artificial refrigeration by using a pump to create a partial vacuum over a container of diethyl ether which, when it boiled, absorbed the surrounding heat, and as a side effect a little ice was formed. Everybody was very impressed by his experiment, but it was a wee bit nippy that day and as everybody wrapped up warm before heading home to light the fires, nobody could think what practical benefit refrigeration could possibly bring.

In the 19th century ice-harvesting became big business and interest in artificial refrigeration remained incidental and none of the few attempts at developing an artificial ice-maker had proved successful. In the very warm climate of Australia however, the problem of keeping food fresh, and of course beer cool, was much more pressing as there was no ice to be had nearby.

A journalist from Dunbartonshire called James Harrison, who had immigrated to Australia and ran a newspaper in Victoria, became convinced that artificial refrigeration could provide the answer. Harrison studied the prototype fridges to date and came up with his own ice-making machine that he began operating in Geelong in 1851. The machine was successful and in 1855 he was granted a patent for an ether liquid-vapour compression refrigerator. By the 1860s a dozen of his icemakers were in operation in Australia, making his the first commercially successful refrigeration system in the world.

The population of Australia in 1850 was only one million while the number of sheep in New South Wales alone was greater than fifteen million, never mind all the cattle, and not even the Aussies could eat that many meat pies. In Britain on the other hand there were meat shortages, and traders from Australia and New Zealand saw an obvious market – if only they could find a way of keeping their produce fit for human consumption during the 8,000-mile (13,000-km) voyage by sea. The first attempt at shipping refrigerated beef from Australia to Britain in 1873 failed when Harrison's cooling system broke down and the rank meat had to be thrown overboard.

Harrison returned to journalism and became editor of a Melbourne broadsheet, *The Age*, which incidentally was run for more than a century by the Syme family, originally from North Berwick. His work on refrigeration had not been forgotten, however, and in 1880 a shipment of frozen beef, lamb and butter arrived successfully in London using Harrison's refrigeration techniques. Refrigerated meat would transform the economies of Australia and New Zealand and within fifty years refrigeration had replaced ice packing as the means of preserving produce in transit and in the process began the conundrum of 'mutton shipped as lamb'.

THE SCHOOL OF IGNEOUS ROCK

72 JAMES HUTTON (1726–97)

The Scottish Enlightenment Gang were a brainy bunch to say the least. David Hume [32] and Adam Smith [23] would often spend an evening in the pub with the man who became known as the Father of Geology, James Hutton. The games of dominoes between the three could last for hours as each player carefully worked out the philosophical, economic and historic consequences of playing the double-four. But while Hume and Smith both looked to the present and the future, Hutton was very much drawn to the past.

An Edinburgh-born farmer, Hutton trained as a doctor but his passion was old rocks. He would spend most of his life tramping around East Lothian, the Borders, the Cairngorms, Arran and in Edinburgh at Salisbury Crags, observing rock formations and taking samples for further investigation. It began to dawn on Hutton that the perceived wisdom as stated in the book of Genesis – that God created the earth around 6,000 years before and nothing had changed since Noah and his menagerie stepped off the Ark and went forth and multiplied – was not consistent with what his investigations were telling him.

Everywhere he went Hutton found examples of land

erosion and far from one standard rock formation, there were layers of differing rock, some of compressed sediment and some volcanic. Hutton went by boat with fellow geologist John Playfair – born in Angus – to Siccar Point on the Berwickshire coast in 1788. The rock layers there are clearly defined and Playfair later commented that, 'The mind seemed to grow giddy by looking so far into the abyss of time'. There was also the possibility that Playfair simply felt giddy because he was on a wee boat in a not inconsiderable swell.

Hutton concluded that the earth was a continually evolving process of erosion and new rock formation that had been taking place very slowly and for a very, very long period of time. Furthermore, the Bible's 6,000-year-old earth seemed to be a major underestimation. Hutton finally published his theory of uniformitarianism, meaning gradual change over a very long period and that all geological changes can be explained by existing processes such as erosion and volcanic action in his literally ground-breaking book, *Theory of The Earth*, in 1785.

His views concerning the earth's formation were, unsurprisingly, attacked by the men of the cloth: if the Bible said the earth was 6,000 years old then who was this Scottish farmer to say otherwise? The real impact of what Hutton put forward in *Theory of the Earth* was not fully recognised until after his death, mainly because the style in which he wrote was said to be so difficult to understand that most people had no idea what he was going on about. It would take the more comprehensible approach taken by his friend Playfair in *Illustrations of the Huttonian Theory of the Earth* in 1802 and further development of his work by Charles Lyell (73) to bring the full depth of Hutton's work to light.

The complexity of his prose was criticised but it was Hutton who, when discussing the age of the earth, came up with the famous line, '. . . we find no vestige of a beginning, no prospect of an end.' Hutton was too canny or too modest to ever attempt to commit himself to an actual age of Earth, but by stating that the planet was much older than had been previously thought Hutton fundamentally changed the very way that we looked at the world that we lived in – although perhaps appropriately in his own lifetime this change would be very gradual indeed.

THE MISSING SCOTTISH LINK

73 CHARLES LYELL (1797–1875)

Scots have always had a great ability to square the circle when points of view conflict: Scottish and British, for example, or the Calvinistic ethos of hard work, thrift and reserve combined with an alcohol intake that is the eighth highest in the world. The Scottish Enlightenment encapsulated such contradictions in that the Church of Scotland's promotion of the importance of education and the search for knowledge increasingly led the world of science into opposition with the world of the Kirk. For David Hume [32] the answer was simple – there was no God (probably), but for most the scientists, philosophers and innovators of the 18th and 19th centuries, their momentous discoveries had to be somehow made compatible with their religious faith.

One man who found this dilemma more difficult than most was the geologist Charles Lyell who was born near Kirriemuir in Angus. Lyell was inspired by the earlier work of fellow countrymen and fellow geologists James Hutton [72] and John Playfair, which had shown the world was considerably older than the 6,000 years stated in the Bible – but theirs was not yet a widely held view. In the 1820s Lyell travelled to France and Italy to study rock formations in the company of geologist Roderick Murchison from Muir of Ord in the Highlands, who would later find fame for defining the Silurian geological period. From 1830–33 Lyell published three volumes called *Principles of Geology* that would prove to be hugely influential and the most comprehensive work on old rocks ever written.

Principles of Geology sets out Lyell's view of uniformitarianism that the world was slowly changing through natural processes and that these same natural processes have been in effect since the world began or as Lyell put it, 'The present is the key to the past'. He also categorised the geological periods of Pliocene (most recent), Miocene (less recent, but still not long ago all things considered) and Eocene (dawn of recent, which sounds like quite recent, but is actually very old), all of which are part of the Tertiary period from two million years old to sixty-five million years old – give or take a million years or so.

When the young English naturalist Charles Darwin set sail on the *Beagle* expedition in 1831 at the start of its five-year journey around the world the ship carried a copy of the first volume of *Principles of Geology*, with the subsequent volumes being shipped out as they were published. Darwin, who as a teenager had studied medicine unsuccessfully for two years at Edinburgh University, was inspired by Lyell and his uniformitarian view and when he arrived in South America he wrote that he saw the world 'through Lyell's eyes' – not knowing that Lyell actually had very poor eyesight. When the *Beagle* finally returned to Britain in 1836 the two men met and became firm friends. Lyell knew of Darwin's theories and repeatedly encouraged and pushed Darwin to publish what would eventually evolve into *On the Origin of Species* in 1859, but although he was always supportive of his friend, Lyell's religious faith and cautious pragmatism prevented him from coming out publicly in favour of all of Darwin's evolutionary theories – when it came to survival of the fittest and humans evolving from apes, Lyell didn't love it.

Within his field Charles Lyell is recognised internationally as a giant, and for anyone fortunate enough to study earth science his work is integral to its understanding. Charles Darwin may have been braver and more radical, but it was only through studying Lyell that he had set out on his revolutionary and evolutionary career. And although Lyell does not have the popular recognition afforded to Darwin, there are two mountains – both over 10,000 feet (3,048 m) high – which bear his name; Mount Lyell in Canada and Mount Lyell in California, which for someone who dedicated his life to the study of rocks, must have been gneiss.

SIZE ISN'T EVERYTHING

74 THOMAS GRAHAM (1805–69)

Born in France of an Irish father and Scottish mother, the scientist Joseph Black had an integral role in both the Scottish Enlightenment and the history of science.

Black studied under William Cullen [71] at Glasgow University and became professor of medicine at Glasgow and then professor of medicine and chemistry at Edinburgh. He was a great friend and drinking buddy of Adam Smith [23], David Hume [32] and James Hutton [72] and was a mentor to a young instrument maker by the name of James Watt [52]. In the early 1750s Black discovered that air was made up of different gases and credited as the first to define carbon dioxide. However, during Black's research he found that when the carbon dioxide was absorbed there was still some air remaining and he decided to let his assistant, Daniel Rutherford from Edinburgh, figure out what this air was. In 1772 Rutherford discovered through various experiments with candles and unfortunate mice that this remaining air could not be absorbed, could not sustain a flame and was not the best atmosphere for mice to live in. Rutherford called the air 'noxious air', but it would later be renamed nitrogen, which it turns out makes up more than seventy-five percent of the Earth's atmosphere and is found in all living things. In view of the omnipresence of nitrogen it is surprising that the name of Daniel Rutherford is not better known today, but then again when you call your greatest discovery 'noxious', historical posterity is perhaps not the first thing on your mind.

While Black and Rutherford were studying the air that we breathe, or in the case of nitrogen that we don't breathe, other Scots were concentrating on what constituted life itself. Robert Brown was a botanist from Montrose who made his name studying the flora of Australia before returning to Britain where he made good use of the new microscope he had got for Christmas. In 1827 while studying plant pollen floating in water he saw minute particles zigzagging in a random fashion, and when he repeated the experiment with dust the same phenomenon occurred. Although he did not know it at the time what Brown was observing was the first physical demonstration of the existence of atoms. It would take another seventy years and Albert Einstein, no less, in the 1900s to explain that, the reason for the random movement of the pollen was that the invisible molecules of water were affecting the visible spores of pollen. This movement would be given the name Brownian motion in honour of the Scottish botanist, as it was felt that calling it a 'Brown motion' would have rather different connotations.

Thomas Graham from Glasgow was also interested in getting to the bottom of all things small. A brilliant chemist who became professor of chemistry at University College, London he was fascinated by the motion of atoms in gases and fluids, and by how different gases reacted when mixed together – which was called diffusion – and the rate at which gases escaped by going through a small hole into a vacuum – which was called effusion. In 1846 Graham published the formula that most people recognise most readily, which put simply stated that, the denser the gas, the slower it will diffuse. This became known as Graham's Law of Diffusion and there was also Graham's lesser-known Second Law, 'Do not touch my test tube'.

Graham is today best known for his work on colloids, with colloids being substances that took a long time to diffuse, and could not be separated by filtering or gravity and made a gluey-like solution; however the particles of these solutions could be dispersed by going through a second substance. Graham gave these non-separating substances the name colloids, from the Greek world meaning 'glue'; and it was also Graham who came up with the terms 'sol' for a solid colloid that was dispersed in a liquid and 'gel' that was a liquid dispersed in a solid. To help him separate substances in solution into colloids and non-colloids, in 1861 he found that a semi permeable material such as parchment was effective for extracting, for example, urea from urine. He called this process 'dialysis', a discovery that would lead in the 20th century to the dialysis of blood and the development of dialysis machines and the treatment of kidney failure.

The word 'sol' in turn would eventually lead to 'aerosol' meaning a solid or a liquid colloid dispersed in a gas and when combined with hair gel it is fair to say that without Thomas Graham the 1980s could never have happened.

Why were so many brilliant Scottish so obsessed with researching the minutiae of science? Was it a subconscious response to the country's longstanding inferiority complex that made scientists north of the border study molecules and atoms? Was it perhaps a realistic approach from a pragmatic nation that prides itself on the phrase, 'many a mickle, makes a muckle'? Or, in Thomas Graham's case, maybe he devoted so much time studying what could be extracted from urine simply because he found himself having to go to the toilet a lot.

IT'S SCOTLAND'S PARAFFIN!

75 JAMES YOUNG (1811–83)

It was only in the 1950s that oil replaced coal (ah, good old coal) as the world's greatest provider of fuel and not that long after, in 1970 and '71 respectively the Forties and Brent oilfields were discovered in the North Sea, with the subsequent forty years coincidentally seeing the rise of Scottish nationalism and the Scottish National Party (SNP) as a viable political movement. From the early '80s the UK was in the very fortunate position of being a net exporter rather than a net importer of oil and was, at the height of its oil production, one of the top ten petroleum producers in the world, and over the past four decades almost every question concerning the merits or otherwise of the Scottish and UK economic models usually ends in a circular argument between Nationalist and Unionist about whether the oil and gas fields of the North Sea belong to Scotland or Britain, and does the Scottish economy subsidise the English, or does the English economy subsidise the Scottish, and whose turn is it to pay for the Welsh.

There is a certain irony for Scotland that the more radical policies of the British politician with whom they were most diametrically opposed, Margaret Thatcher, were only economically possible due to the funds flowing into the Treasury coffers from North Sea Oil, and the slogan that made the SNP's name in the 1970s, It's Scotland's Oil, still resonates today, although granted on a reduced level as oil production has declined, with perhaps the ultimate irony being that if and when Scotland becomes an independent nation, the wells will finally run dry.

It will be a sad day when the whales of the North Sea, if there still are whales in the North Sea, watch the last oil tanker sail away, especially when you consider that the very first oil refinery in the world was opened in Scotland in 1851. The man who opened it was a chemist by the name of James Young who hailed from Glasgow. Young studied chemistry at evening classes at what is now Strathclyde University where he was taught by Thomas Graham [74] and became life-long friends with David Livingstone [47], whose expeditions to Africa, Young, in later life, would help fund.

After completing his studies Young was working as an industrial chemist in England when, in 1848, a friend told him about a seam of oil that was seeping from a colliery in Derbyshire. Unrefined mineral oil had been around forever, but when Young distilled the Derbyshire oil he found that it could be used for lighting lamps, which had formerly been lit by whale oil. Convinced that this oil must have come from coal, he helped set up a company and began to experiment on different forms of coal until, in 1850, by distilling and heating cannel coal from West Lothian he found that it produced oil that would burn easily and produce a bright light.

Young immediately saw the commercial opportunity and relocated to Bathgate in West Lothian where he opened an oil works to produce and sell this new refined oil. He called it paraffin oil, as when the oil congealed it resembled paraffin wax, and he called the process of breaking up the oil into different substances, 'cracking', a term still used in oil refining today. When Young exhausted the local supplies of cannel coal in 1866, he turned to the more readily available local shale to produce his paraffin. As a result, shale oil became one of the major industries in Scotland in the late 19th century and Young's paraffin and paraffin lamps were exported around the world. The success of his discovery would give the chemist the nickname James 'Paraffin' Young and make him an extremely wealthy man.

It has to be said that James Young was not the only person burning the midnight oil to produce oil that could be burned at midnight. In 1846 Canadian Abraham Gesner demonstrated how you produce oil from coal and called it kerosene, which remains the name used in America today. However, it was Young who was the first to patent, both in Britain and the US, and it was Young who was the first to open a refinery, eight years ahead of the first oil well to be drilled in the United States in 1859 – although it has been said that while he was a very successful businessman, if he had taken up some of the opportunities offered to him in the States it might have been the Youngs rather than the Ewings who became America's best known oil barons of Scottish origin.

In the 20th century paraffin proved no match for the more readily available refined petroleum as an energy provider, but because it is less volatile paraffin is still commonly used all around the world for lighting, heating, fuel and in cosmetics. And if that is not enough of a legacy to be going on with, then the advent of paraffin also saw the phasing out of whale oil as a fuel for lighting. So even if the shale industry in Scotland may be no more at least we still at least have the occasional whale.

LIVING IN THEORETICAL ELECTROMAGNETIC DREAMS

76 JAMES CLERK MAXWELL (1831–79)

If you stroll eastward along George Street, the second street of Edinburgh's New Town, and after you have passed all the designer stores and upmarket bars on your way to a little light retail therapy in Harvey Nichols, you will pass by a statue of a bearded man wearing disconcertingly ill-fitting clothes; he is seated, holding what looks like an artist's palette and with a wee dog at his feet. On closer inspection one finds that the man in question is neither an artist nor Greyfriars Bobby's dog-walker, but is in fact James Clerk Maxwell – Scotland's greatest ever scientist and alongside Isaac Newton and Albert Einstein ranked as one of the three greatest scientists the world has ever known. Maxwell is not as well known today as are Alexander Graham Bell [96] or John Logie Baird [99], as he did not change the world by inventing something. But Maxwell's reputation is based on his work as a physicist and mathematician that described in theoretical terms the very way that we see the world we live in and he would be called the Man Who Changed Everything. His dog was called Toby.

James Clerk Maxwell was born in Edinburgh, but moved with his family to the countryside near Dumfries. From a very early age he showed an inquisitive nature, but unlike most curious children who are happy to endlessly ask, 'Why?', Maxwell also wanted to know 'How?', and would not be satisfied until he had an answer. Unsurprisingly his family soon got tired of this and when he was eight years old sent young James back to Edinburgh where the poor teachers at the Edinburgh Academy had to contend with his thirst for knowledge. Because of his country accent and clothes he was nicknamed Daftie by his fellow enlightened students and it would be a while before he began to show his true academic potential.

Maxwell went to the universities of Edinburgh and Cambridge and when he was only twenty-five he was appointed as a professor at Aberdeen University in 1856, before going on to fill a similar position at King's College, London, before finally returning to Cambridge. He had a keen interest in astronomy, and in 1859 he explained one of the great celestial mysteries by stating that the rings of Saturn were not in fact solid but millions of small particles orbiting the planet. The scientific contribution for which he is most famous, however, came from his research much closer to home.

In the mid-19th century the understanding of electricity was in its infancy. It had been as recently as 1831 that English physicist Michael Faraday and, independently, the American son of Scottish immigrants, Joseph Henry, first discovered electromagnetic induction. Between 1861–73 James Clerk Maxwell published a series of comprehensive works on this new technology which up until that point nobody had completely understood, culminating in his 1873 *Treatise on Electricity and Magnetism* and his series of four mathematical equations – for these, he drew on the best of all the previous studies by Faraday and others and added his own research and theories.

Maxwell had a genius for being able to understand instinctively how things worked and would then concentrate on the maths and the physics to back his instincts up. He theorised that electricity and magnetism were actually oscillating waves that travelled through space and created electromagnetic fields, and that these waves travelled at the speed of light and could be of differing frequencies, that light itself was an electromagnetic wave which happened to be visible to the human eye and that radio waves existed. The four equations became known as Maxwell's equations and became the basis for the science of electromagnetism.

Maxwell was aged only forty-eight when he died in 1879, and it would be left to others over the following decades to prove that his theories on electricity, light and radio waves were in fact correct. And because he was right in so many aspects, it is possible to say that without his contribution to electromagnetism the invention of television, radio, microwaves, radar and mobile phones would not have been possible. Albert Einstein, who kept a photo of Maxwell in his study, described him as the greatest physicist since Newton and stated that his own Theory of Relativity would not have been possible without Maxwell's theories of electromagnetism and, moreover, that Maxwell had, 'changed our conception of reality'. Not bad for somebody who used to be called Daftie.

But what is James Clerk Maxwell holding in that statue on George Street? It is not in fact an artist's palette, but it is instead what he called a colour top or colour wheel. Maxwell had studied colour and colour blindness and discovered that by mixing the primary colours of red, green and blue you could produce the colour white. Emboldened by this he decided to take it one step further. The world's first photograph had been taken in 1825, and so far all photos were in black and white, but in 1861 in London Maxwell decided to bring a little colour into the world. He asked a photographer to take a picture of the same object three times, but when the photos were taken they were done so through three, different coloured filters (blue, red and green). Then, in front of an invited audience the three exposed slides were projected through three different coloured lanterns (again blue, red and green) onto one screen at the same time. When the slides were brought into focus, the other colours in the photograph, including white, were clearly evident and the result was the world's first ever colour photograph. Giving the demonstration a patriotic touch the object in question was a tartan ribbon, but to the disappointment of the Scottish tourist industry it would not be until the 1930s that colour photography began to become popular as it had quickly become clear that the world was not yet ready for tartans that included yellow and brown.

EVEN SOME SCOTS LIKE IT WARMISH

77 JAMES DEWAR (1842–1923)

Scotland has a long and illustrious history of investigating the cold things in life – as in the case of William Cullen [71], for example, with artificial refrigeration in 1756 – but, sometimes, even the most Calvinist of Scots have to admit that when you are cold and tired and weary and a long way from home, nothing quite beats a drop of tepid tea with a slightly metallic aftertaste. And true enough you will find that a Scot invented the vacuum flask, although interestingly not as a method of keeping liquid warm, but as yet another way of keeping things as cold as possible.

James Dewar was born in Fife and studied chemistry at Edinburgh University before moving south in 1875 where he became a professor at Cambridge and then the Royal Institution in London. Dewar had an illustrious career studying and researching many aspects of physics and chemistry, but became best known for his work in the liquefaction of gases and low-temperature technology. In 1898 Dewar became the first person to liquefy hydrogen – which would later be used as rocket fuel – and, the following year, to solidify hydrogen. It was not, however, the discovery of liquid hydrogen for which he would make his name, but for the machine that he used to store the liquid gases in.

Dewar had been working on liquid oxygen since the late-1870s and by the early-1890s he was producing large amounts of it in his London laboratory, but he needed to find a way to keep the liquid oxygen cool. He came up with a double-lined vessel with a vacuum between the layers that prevented convection and conduction of heat, and it had silver interior walls to prevent heat transference by radiation. Not that this was the reason that Dewar made his flask, but as well as keeping the liquid cold, it also meant that liquid could be kept warm. However Dewar was a scientist and not a businessman, and did not patent his invention or foresee any other potential for what he had done. It was not until 1904 that two German glass blowers discovered that Dewar's flask was ideal for keeping milk warm. They began to produce the flask commercially and renamed it Thermos after the Greek word *therme*, meaning heat.

James Dewar was known for being a prickly and sometimes difficult character, who preferred working on his own rather than socialising with colleagues. Dewar by name and Dewar by nature. When he discovered what the Germans had done with his invention, he was predictably, if understandably, outraged by this development and sued the Germans, but because he had never patented his invention he was ultimately unsuccessful and powerless to stop production. There had never been any question of the Germans or anyone else denying that Dewar was the inventor of the vacuum flask, but Dewar never received any financial reward for its commercial success and, over time, the term 'thermos flask' would replace the term 'Dewar flask'. A disgruntled and angry Dewar remained in his laboratory with his refrigerated liquid gases and vowed never to go on a picnic again, or anywhere where there might be one of those damned flasks, much to the amusement of his much put-upon work colleagues who began bringing in flasks of tea just to wind him up.

There was one other famous invention for which James Dewar was responsible, or to be accurate co-responsible for, and that was cordite – a smokeless British replacement for gunpowder. It was made up of nitrocellulose and nitro-glycerine and was invented by Dewar and Frederick Abel in 1891. The British in both World Wars used cordite extensively, and in 1916 a huge cordite factory was built at Gretna on the Scottish-English border; the factory employed more than 16,000 workers, 11,000 of which were women. Gretna is of course famous for being a location where young, and not so young, couples go to get married, and with such an influx of young workers into the town there were inevitably a considerable number of weddings, but working in munitions as they did at least they had plenty of experience of 'keeping their powder dry' until the happy day.

FEELING A LITTLE GASSY

78 WILLIAM RAMSAY (1852–1916)

When in 1938, *Superman* creators Jerry Siegel and Joe Shuster were looking for the name of the planet that the baby Kar-El escaped from moments before its destruction, they decided that Krypton, the name of a gas that had only been discovered forty years previously sounded suitably alien. And the man who discovered krypton was a Scottish chemist from Glasgow who also discovered argon, neon and xenon – which to be honest all would have perfectly acceptable alternatives.

William Ramsay was born in Glasgow and studied chemistry at Glasgow University. He was appointed professor of chemistry in Bristol in 1880 and moved to the prestigious University College, London in 1887. By the late-19th century scientists had worked out that the earth's atmosphere was made up of around 78% nitrogen, 20% oxygen and a smidgeon of carbon dioxide, but this only came to around 99%. What was the last one percent made up of?

In 1894 when collaborating with English scientist Lord Rayleigh, Ramsay discovered a colourless, odourless gas that they named argon, which was Greek for 'inert'. A year later Ramsay became the first person to discover another gas, helium, which scientists were aware of but had not until this point isolated – although it took Ramsay and his assistants a few days to announce their discovery as they were enjoying talking in high-pitched voices so much. Helium has the lowest boiling and melting point of any element and is colourless, odourless and non-flammable. It has become popular for use in balloons, although it is so light that it would take around 6,000 helium balloons to lift a child into the air – one of those interesting facts that makes you wonder why they wanted the unfortunate youngster to be airborne in the first place.

Even with the discoveries of argon and helium there was a tiny bit of the atmosphere still to be discovered, and over the next three years Ramsay continued his painstaking work to find out what was there. Ramsay would cool air until it was liquid and then slowly heat this liquid air and isolate the different gases, and in 1898 he found three further gases in tiny quantities that he named neon (Greek for 'new'), krypton (Greek for 'hidden') and xenon (Greek for 'stranger'). Of the three it would be neon that would have the greatest impact as it was found that it gives off a luminous red glow when electricity is conducted through it. In 1923 the first neon signs were used for advertising in the US and over the following decades all over the world hotels, theatres, bars, cinemas, launderettes and other establishments where clothes are removed would be lit with neon signs.

Helium, argon, neon, krypton, xenon plus radon became the noble gases, and appropriately William Ramsay would in 1904 become the first Scottish recipient of a Nobel Prize for his achievements in the field of chemistry. Sadly as krypton has exceedingly low chemical reactivity it has not been possible as yet for scientists to discover green kryptonite, or for that matter any colour of kryptonite, but Scotland still remains very proud of their scientist who discovered those extra gases and in celebration Scottish brewers have for years been adding a little extra gas to their lager in his honour.

IT'S A SMALL WORLD, BUT I WOULDN'T WANT TO IRRADIATE IT

79 CHARLES THOMSON REES WILSON (1869–1959)

It was the father of nuclear science, the splitter of the atom and son of a Scotsman, Ernest Rutherford, who described the cloud chamber as the 'most original and wonderful invention in scientific history'. Invented in 1911, the cloud chamber allowed the movement of energised particles to be detected and proved integral to the history of 20th-century nuclear science, and it all came about because an ex-pat Scot liked nothing better than to climb Scotland's highest peak when he was back home.

Charles Thomson Rees Wilson, or C. T. R. Wilson for shorter if not exactly short, was born on a Midlothian farm just south of Edinburgh, but moved at the age of four to Manchester with his mother when his father died. Thomson was an academic child and would have an excellent university career studying science and graduating from both Manchester and Cambridge universities. It was at Cambridge that he began to specialise in meteorology and it was this interest in the weather that brought him back to Scotland in 1894. It was the weather, too, that enticed him up Britain's highest mountain, Ben Nevis, which from 1883–1904 had a permanent observatory located on its summit.

Wilson was fascinated by the different cloud patterns and optical phenomena that he saw from the top of Britain and was determined to recreate them back in his Cambridge laboratory. In 1880, Falkirk-born physicist John Aitken had invented the most advanced meteorological chamber and it had yet to be bettered; it was a glass jar in which clouds could be formed artificially if water-saturated air in the glass contained dust, but Aitken had not been able to produce clouds from air that was dust-free. Through experimenting and expanding the amount of air in the jar Wilson was able to produce water droplets and therefore clouds, even though all dust had been removed.

This led Wilson to theorise that there must be something in the air, such as charged particles or ions, for the droplets to attach themselves to. He then experimented with the new-fangled invention of X-ray – invented in 1895 by German physicist Wilhelm Roentgen – and found that even larger amount of droplets were produced when radiation was added to the air. He continued to experiment until in 1911 he was able to produce a photograph of one of his cloud chambers. Now, for the first time, rather than scientists having to hypothesise and theorise about the subatomic world, everybody could see these charged particles for themselves.

Charles Thomson Rees Wilson continued his eminent career in physics and meteorology before retiring to the Midlothian countryside of his childhood. In his long life he received many honours culminating in the Nobel Prize for Physics that he was awarded in 1927. It was thirty-three years from C. T. R. Wilson beginning his experiments to recreate the cloud patterns that he had witnessed on Scotland's highest mountains to receiving the Nobel Prize, and 16 years since he had finally achieved his goal, but as all serious hill-walkers will tell you it's not just the getting to the summit that counts, you have also to savour your Scotch egg and cup of tea before safely getting down.

I WILL ALWAYS LOVE EWES

80 DOLLY THE SHEEP (1996–2003)

World-changing innovations and inventions are, mostly, the result of many years of painstaking research and investigation. The instigator – be they scientist, engineer, philosopher, whoever – might have worked with a colleague, assistant or technician, but history tends to remember the inventor, and the poor lab assistant who made the tea is sadly forgotten. By the end of the 20th century it became increasingly accepted that knowing who takes two sugars and who likes which biscuit was a crucial part of the process and it was a multi-national team, rather than one individual, which is credited for one of the most groundbreaking discoveries in science.

The perception of your archetypal scientist is usually one of a bespectacled, white-coated, slightly eccentric, obsessive who spends every waking minute in his laboratory. However, when the Roslin Institute team in 1997 announced that they had achieved the world's first successful cloning of a mammal from an adult cell we discovered that even the most cerebral of scientists were not all that different from the rest of us. With the cloning process a success, but only known by the research team, and its announcement to the world therefore imminent the team meet to decide on what the animal – a sheep – should be called. Should they go for a classical Greek word, something that implies 'new life' or 'rebirth'? Or something that refers to a famous scientist from the past? Or should they, perhaps, name their clone something reminiscent of one of the team-members? All these possibilities receive serious, detailed discussion before someone suggests in passing that, as the clone's cell came from an udder, the name might relate to that. Picture the scene as all these brilliant embryologists discuss the relative merits of various famous mammary glands – whether enhancements should be allowed and does size matter – before finally and perhaps predictably plumping for the largest breasts they can think of. And with the historic decision finally made the team who created Dolly the Sheep would finish work for the day.

The Midlothian village of Roslin has a population of fewer than 2,000 yet a significant number of claims to international fame. For a start, it was the birthplace of the inventor of Bovril, John Lawson Johnston, and it is home to one of Scotland's most fascinating buildings, the 15th-century Rosslyn Chapel which, as a result of the many mysteries that may or may not lie therein, featured in Dan Brown's novel, *The Da Vinci Code*. There has, also, been since the 1940s an agricultural research and animal science facility in one form or another based close to the village and this was renamed the Roslin Institute in 1993. English-born embryologist Ian Wilmut had been working at Roslin since 1974 and he, alongside fellow Englishman Keith Campbell and their assembled team began the research in 1991 that put Roslin centre stage, again, and long before anybody had heard of *The Da Vinci Code* and the controversial theory that the Holy Grail did in fact contain a warm salty beef liquid.

A clone is an organism which is genetically identical to its parent and it is the result of asexual reproduction – which means that no matter how spookily similar identical twins may appear, they are not in fact clones. Cloning does however occur in nature; many plants, for example, reproduce asexually and man has for thousands of years used cloning in horticulture by taking cuttings or making grafts to preserve or increase the numbers of individual plant varieties. In 1952 two American scientists, Robert Briggs and Thomas King, were the first to successfully clone an animal, albeit an amphibian, a tadpole to be precise, and they achieved it by a process known as nuclear transfer: the nucleus that contains DNA is removed from a non-reproductive cell and implanted into a reproductive cell which has had its nucleus removed. However, this did not lead to the major breakthrough in cloning that many had expected and for the next forty years the predictions of science-fiction writers remained on hold.

In 1996, after 277 attempts, the Roslin team finally succeeded in extracting unfertilised eggs from the DNA cells of a six-year-old adult sheep. They implanted the eggs into thirteen ewes and on 5 July, one ewe gave birth to a lamb, Dolly, the world's first cloned mammal created from an adult cell. Dolly's surrogate mother was a Scottish Blackface, Britain's most popular breed, but Dolly was a politically correct Finn Dorset.

When the Roslin team announced their achievement in 1997 Dolly became an overnight celebrity, but for security reasons she never lived outside her enclosure in the Institute and was never to go outdoors or ever see islands in the stream. This did not, however, stop her giving birth six times, with her first lamb, Bonnie, being born in April 1998. Her lambs were conceived in the normal manner and although she had the strangest feeling that she had enjoyed better in the past, Dolly was quick to assert herself when a rival ewe called Jolene looked as if she might take her ram. Sadly, by the age of five – or was it eleven? – Dolly began to show signs of premature ageing and developed arthritis, and to prove that deep down she was a very Scottish sheep after all, she died of arthritis and lung disease in 2003. Dolly's life ended on St Valentine's Day, although the Roslin Institute were not completely without feeling as before they put her to sleep they allowed her to wear her blonde wig one final time.

Entertaining Scots

The feedback from tourists to Scotland is, in general, that they enjoy the experience. They find the people hospitable, the history fascinating, the landscape stunning and the information centres staffed by extremely helpful, knowledgeable and multilingual Spaniards and Poles. On the whole, visitors' expectations are surpassed and even the weather is not usually as god-awful as they had imagined, but tourists do tend to mention how surprised they were when, having asked where they might find some traditional Scottish entertainment, they ended up in a karaoke bar singing along to *Don't Stop Believing* by Journey.

Yet, while a large faction of Scots eschew Dashing

White Sergeants in favour of salsa evening classes, Scotland does have a rich history of which it is rightly proud in the fields of sport and the arts and entertainment industry; and, anyway, they are only going to the salsa classes until such time as they actually find a boyfriend.

NEVER PLAY GOLF WITH THE DUKE OF CUMBERLAND, HE PROBABLY CHEATS

81 DUNCAN FORBES (1685–1747)

Golf and whisky. Whisky and golf. If and when all the Scottish financial institutions in Edinburgh either go bust, close down or head south, if and when all the oil runs out and if and when the bodies of a plesiosaur family are found at the bottom of a large Highland loch, then we can guarantee that Unionists will be gleefully telling the nationalists that they were right all along – Scotland really is rubbish and unworthy and should never be allowed out without parental supervision. And the beleaguered, despondent, defiant nationalists will respond in unison with the cry, 'Golf and whisky, whisky and golf'. For as long as there is barley and water and grass and sand, there will always be at least two Scottish inventions exported around the world, and both of which bring visitors to its shores no matter what the exchange rate is.

The origins of golf are lost somewhere in medieval times. There is speculation that the game may have been imported from the Netherlands as game called 'kolf', meaning 'stick', and others have suggested even earlier examples of games in China and India that involved sticks and stone-like balls, but for all intents and purposes the game of golf as we know it was invented in Scotland.

The fact that golf, or 'gowff', has been around for well over 500 years is demonstrated by a 1457 Scottish Act of Parliament which prohibited the playing of the game as Scottish nobles were spending too much time on the 'gowff' course when they should have been at archery practice and were showing worrying signs of hooking and slicing their arrows. James II was the Stewart king responsible for golf's prohibition, but other Stewarts/Stuarts were enthusiastic participants and promoters of the game. James IV's official accounts showed that he bought golf clubs in Perth in 1502. In 1641 on one of his rare trips to the land of his birth, Charles I [12] was playing golf at Leith when he was told of the Irish Rebellion, but true to form decided to ignore this bad news in favour of finishing his round. In 1681, also at Leith Links, the Duke of York, who would briefly become James VII of Scotland and James II of England, partnered local man John Paterson against two English noblemen in the first recorded international golf match. And we should not forget that Mary, Queen of Scots [3] was said to be a keen golfer. She was reported to be playing in Musselburgh in 1567 only a day or two after the murder of her husband, Lord Darnley, in which many believed Mary was complicit, but whether she was or not, at least we can safely say that Mary, Queen of Scots was the world's very first golfing widow.

The heartlands of golf have always been the coastal areas of Lothian and Fife, in the east of Scotland, where there is plenty of gently undulating sandy ground ideal for links courses. Golf has been played at the legendary Old Course St Andrews since the 15th century and in 1759 the course became the first in the world to have eighteen holes. Five years earlier the Society of St Andrews Golfers had been founded; it was renamed the Royal and Ancient Golf Club of St Andrews in 1832 and, over the years, became golf's governing body and responsible for establishing the rules and organisation of the game. However, no matter how ancient St Andrews is, the world's oldest golf cub is the Honourable Company of Edinburgh Golfers, now long-established at Muirfield in East Lothian but which formed originally in Leith – Leith Links – in 1744 and in the same year published the first ever *Rules of Play.*

The Edinburgh Golfers felt it would be beneficial to have some legal input when drafting the rules and as Scotland's leading judge featured among their number he was given the job, which if nothing else saved on legal expenses. Duncan Forbes was born near Inverness; he was a prominent MP and in 1737 was appointed head of the Scottish judiciary, as Lord Culloden, Lord President of the Court of Session. The Forbes family owned the land at Culloden and when the 1745 Jacobite Rising began Duncan Forbes returned north and played a crucial role in keeping Inverness loyal to the British Crown. However, Forbes could not have expected that the conflict would end within a mile of his family home, with the final battle taking place on Culloden Moor in April 1746. Forbes was a Unionist through and through and had spent his political career opposing Jacobism, but as a Highlander, a judge and a believer in fair play he was appalled by the treatment meted out to the defeated, wounded men in the aftermath of battle and had tried to intercede on their behalf, but had been rebuffed, leading Forbes to conclude that George II and his son the Duke of Cumberland may well be royalty, but they were clearly no gentleman golfers.

Back in the happier days of 1744, Duncan Forbes and the Edinburgh Golfers had come up with thirteen rules of golf in their *Rules of Play* and many remain, unaltered, to this day – rules 4, 5, 8, 10, 11 and 12, should you wish to look them up. However, within the thirteen original rules there are some interesting insights into golfing life in 18th-century Leith. In rule four, for example, which states that you cannot remove any obstructions from the ball, the obstructions listed include stones and, rather more worryingly, bones. Rule five says that you lose a stroke if your ball goes into water or, indeed, if your ball goes into the 'watery filth' and in rule ten – the ball must be played where it lies – reasons for influencing where the ball might eventually lie are given as, 'person, horse, dog or anything else', showing what a busy place Leith Links was in 1744 and demonstrating Forbes's legal background in attempting to cover all eventualities.

IT STARTED UP IN FIFE

82 JOHN REID (1840–1916)

The popularity of golf spread slowly, but inexorably from the 17th century. The court of James VI [11] brought the game to England in 1603 and the oldest club outside Scotland was founded in Blackheath, London in 1766. The first ever Open Championship, which was played over thirty-six holes, was hosted in 1860 at the Prestwick Golf Club in Ayrshire with Willie Park from Musselburgh

– the very first winner of a golf 'major' – finishing two strokes ahead of seven other Scottish professionals.

Golf had been played sporadically in North America during the 18th century with a club formed in Montreal, Canada in 1873, and around the same era clubs sprang up in India, Europe and Australia. The first golf club in the United States, however, was not formed until 1888. John Reid and Robert Lockhart, both emigrants from Dunfermline, had settled in New York; Lockhart worked in the linen industry and Reid managed an ironworks. When Lockhart returned from a business trip to Scotland with a set of golf clubs, the two men realised that there was no obvious place for them to play and initial attempts in Central Park proved unsuccessful as the joggers kept complaining.

Undeterred, Reid devised a three-hole course on pasture land near his home in Yonkers, then, before even the cows became fed up with the quality of the greens, he and Lockhart moved again, creating a six-hole course in the middle of an orchard and the two Scots and their growing band of golfing buddies named themselves, the Apple Tree Gang. The popularity of the games became such that they established a still larger course and set up their own club which, unbeknown to them of course, would become the oldest surviving golf club in the United States. Reid and Lockhart named their proud venture the St Andrew's Golf Club, in recognition of their homeland but with an additional apostrophe to differentiate it from the home of golf, and John Reid was elected the club's first president. St Andrew's became one of the five clubs that together founded the United States Golf Association in 1894 and, at one stage, had fellow son of Dunfermline, Andrew Carnegie [26], as a member – always useful to have the richest man in the world in the club when you're having a raffle.

The growth of golf in America was truly staggering, especially when you consider America's reluctance to embrace any sport they have not invented themselves. From zero courses in 1887, there were more than 1,000 by 1900, and more than 3,000 by 1920. However, for all the popularity of this new game the Americans knew neither how to play it nor how to build and maintain a golf course and, therefore, the new clubs employed hundreds of Scottish golfers from Fife, Lothian and Ayrshire as coaches, course designers, green-keepers and club professionals – and no matter if these Scots were not all champions, they would still be far better than the Americans they were teaching.

The first US Open for professionals was held in 1895 and the following year James Foulis from St Andrews became the first Scot to win the trophy. From 1896–1910 Scottish golfers won twelve out of fifteen championships;

Alec Ross from Dornoch in the Highlands won in 1907 and his brother, Donald, also from Dornoch, designed some of America's greatest golf courses including Pinehurst in North Carolina, Oakland Hills in Michigan and Oak Hill in New York. The greatest of all the early Scottish golfers in the United States was Willie Anderson from North Berwick in East Lothian who between 1901 and 1905 won four US Opens; the only golfer in history to have won three US Opens in succession – a forgotten Scottish sporting hero at home, but still alongside the legendary figures of Bobby Jones, Ben Hogan and Jack Nicklaus one of only four men to have won four US Opens.

Tommy Armour from Edinburgh won the US Open in 1927 and is the last Scot to have done so to date. He also won the USPGA in 1930 and the British Open in 1931 making him the only Scot to have won three of golf's four majors and thus, you could argue, the nation's greatest golfer to date.

It was remarkable that Armour was able to play golf at all, never mind win championships: when he was serving as an officer in the British Army in World War I he was in a mustard gas explosion which rendered him permanently blind in his left eye, temporarily blind in his right eye and with head and arm injuries from which he ended up with metal plates in his head and left arm.

Tommy Armour was much loved by American sports fans who nicknamed him the Silver Scot. But even the greats can have an off day and, interestingly, one week after winning the US Open in 1927, Armour achieved the worst professional score on one hole in the history of golf when he took 23 shots on the 17th, although to be fair he was rather unfortunate that his putt for a 22 lipped out.

STANDING ROOM ONLY

83 ARCHIBALD LEITCH (1865–1939)

The first recognised international match in the history of the most popular team sport in the world, association football, took place in November 1872 between England and Scotland at Hamilton Crescent, Partick, Glasgow, where 4,000 spectators saw either a tactically astute defensive master class by two evenly matched teams, or a truly dreadful game of football as the game ended 0-0. The ground where the game was played was that of a Glasgow cricket team, and many of the spectators were no doubt disappointed that WG Grace wasn't playing. And it is right and proper to mention the fact that the great and noble sport of cricket has an even longer history in Scotland than football does. However, despite this long history, cricket in Scotland is generally perceived and derided as an English game, which is ironic as Scotland's most popular sport, association football, is very much an English game, but with the subtle difference that with football, unlike cricket, the English initially let the Scots take over.

The Football League, the oldest football league in the world, was established in England in 1888. It was the brainchild of Perthshire-born William McGregor, who as a director of Aston Villa FC, was fed up with ad-hoc fixtures and cancelled games. Twelve professional clubs, including Aston Villa, Blackburn, Bolton, Burnley, Everton, Stoke and Wolves, agreed to join; McGregor was chairman, and they called themselves the Football League rather than the English League in the so far fruitless hope that one day Celtic and Rangers might sign up. The League's first winners were Preston North End who went through the season undefeated in no small part because the majority of the team were Scottish. Even with the formation of the Scottish League in 1890, it was players from Scotland who continued to dominate the early years of the Football League, and they were nicknamed Scotch Professors on the basis of their more educated passing game rather than any excessive drinking habits.

As football became more and more popular in Britain it became imperative that stadiums would have to be built to cope with demand. By 1903 and the building of the new Hampden Park in Glasgow, attendance at one match hit 100,000 for the first time, not forgetting the additional thousands who hadn't managed to leave the pub. In 1937, 149,415 supporters watched the Scotland-England international at Hampden – 149,000 Scottish fans and 415 English in disguise – and, until the building of the Maracana in Rio de Janeiro in 1950, Hampden Park remained the largest football stadium in the world, as well as being the largest communal urinal in the world.

The architect who built the new Hampden Park was Glasgow-born Archibald Leitch, Archie Leitch for short. Archie Leach also happens to be the real name of Bristol-born Cary Grant, and while Archie Leach from Bristol went on to make classic films such as *Arsenic and Old Lace*, Archie Leitch from Glasgow went on to make classic football stadiums for Arsenal and at Old Trafford.

Leitch's career as an architect of football stadia might have been over almost as quickly as it began when his first design, Ibrox Park in Glasgow, which was completed in 1900, became the world's first sports-stadium disaster: in 1902 a section of wooden terracing collapsed and twenty-five people died. Leitch, however, was not held responsible for this tragedy and was commissioned to design Hampden Park and in 1929 was also responsible for the new main stands at Celtic Park.

And it was not only in Scotland that the architect was in demand. In the early part of the 20th century when throughout Britain football became the 'people's game' and replaced religion as the opium of the masses, it was Leitch who was given the job of building the new cathedrals where worship would begin every second Saturday at three o' clock, and the Lord's name would often be called upon in reference to a decision by the referee and holy wine would be superseded by holy Bovril, that had been invented by John Lawson Johnston from Roslin in Midlothian, while living in Canada in 1874.

Leitch was responsible for the design and redevelopment of more than twenty major football stadiums in England and, overall, has more iconic buildings to his name than Robert Adam and Charles Rennie Mackintosh [98] combined. He built the main stands at Anfield, Goodison Park and White Hart Lane – the homes of Liverpool, Everton and Tottenham Hotspur respectively. He was responsible for the redevelopment of Villa Park, Aston Villa's ground, and Stamford Bridge for Chelsea. In 1913 he was the architect for Arsenal FC's brand new London stadium, Highbury, named after the local area and whose home it remained until 2006 when Arsenal moved to The Emirates, which was not named after the local area. And Leitch was the architect for the new home of Manchester United, opened in 1910 and named Old Trafford, or the Theatre of Dreams for short. The very first game to be played at the new stadium was a rather tasty seven-goal thriller between Manchester United and the team who would become their greatest rivals, which ended with Liverpool defeating the home team 4-3 – although Alex Ferguson [89] was absolutely furious at the lack of injury time played.

THIS IS STONEHAVEN

84 JOHN REITH (1889–1971)

The BBC is the largest broadcaster in the world; BBC World News the most watched news channel and the BBC World Service has more than 180 million listeners in thirty different languages. The Corporation has seen considerable change and controversy since its inception in 1927, but it remains the oldest public broadcaster in the world and has retained throughout its original mission statement to '. . . inform, educate and entertain', with the notable exception of almost everything commissioned by BBC3.

The man who came up with that mission statement

and the corporation's first managing director was John Reith, the Stonehaven-born son of a Presbyterian minister. Reith had no broadcasting experience whatsoever when he applied in 1922 to be general manager of a new, experimental radio company, the British Broadcasting Company, but with radio in its infancy lack of experience did not prove a hindrance at his interview. Reith had been educated in Glasgow, served in the army during World War I and had worked as a manager at a Glasgow engineering firm. He was a driven, dictatorial man who was determined that he knew best and that he would make this new-fangled wireless a success.

Radio broadcasting had begun in the US in 1920 and the British Broadcasting Company was set up by the Post Office as the first radio broadcaster on home shores. Experimental broadcasts began in London in November 1922 and within five years or so, transmitters and relay stations had been installed throughout the country. The company was privately run and by 1926 more than two million British households had bought a receiving licence. The chimes of Big Ben were first broadcast in 1923 with the Greenwich time signal heard for the first time in 1924. Another first aired in 1923 was the Met Office weather forecast, which proved so popular that the same forecast was then broadcast for the next fifty years, as nobody seemed to mind too much that it was not especially accurate. However, it was the British General Strike in 1926 that established the BBC's reputation for impartiality; newspapers were strikebound and radio, therefore, was the only medium through which the wider public could find out what was going on.

Reith found himself in the unenviable position of being pressurised by the Government on one hand and the Trade Unions on the other, each keen to have their own point of view broadcast. Despite considerable criticism Reith managed to establish the reputation that the BBC has become known for throughout the world – being stuck in the middle and pleasing nobody.

In 1927 with radio firmly established countrywide, the British Broadcasting Company became the British Broadcasting Corporation (BBC) and the world's first national public service broadcaster. It was established by the government and paid for through the licence fee, but would remain independent with regard to its broadcasting content. Reith was appointed managing director and became more dictatorial still: running the corporation as his personal fiefdom, ensuring that standards were upheld, that dinner jackets were worn at all times and yet, crucially, despite his unbending Presbyterianism, he made sure there was enough music, sport and entertainment broadcast so that people actually tuned in.

Since 1927 the BBC's motto has been, 'Nation Shall Speak Peace Unto Nation', but it was not until 1932 that they did anything about it with the first short-wave broadcast of the Empire Service, the forerunner of the World Service. Reith was somewhat pessimistic about the new Empire Service, having said after the initial broadcast that the programmes would, 'neither be very interesting or very good'. But despite this vote of no confidence the World Service did manage to find an audience and in 1938 the BBC began providing first an Arabic- and then a German-language option, and by the time of World War II was broadcasting throughout Europe, where the words, 'This is London', came to mean far more than finding out the result of the Man United game.

Reith was also less than enthralled by the new medium of television, invented by his former fellow classmate at Glasgow Technical College, John Logie Baird [99], and remained consistently apathetic toward its potential. Eventually he agreed to experimental television broadcasts in 1932, but continued doggedly to put as many obstacles in his fellow Scot's way as possible and when the first, official BBC television broadcast was made in November 1936 it was the rival Marconi television system which screened it. Curmudgeonly to the end, Reith declined to appear in the first television transmission, broadcast in August 1936, and it is believed that 500 households witnessed this moment of broadcasting history, a modest audience, true, but one that has never done BBC Alba any harm.

If Reith was uninterested in television, he was more than happy to be the public voice of radio and in so doing became a figure of international renown. When Edward VIII made his abdication speech to the nation and to the world, he did so on BBC Radio, but it was Reith's introduction of him, not as King but as Prince Edward, which signalled to every listener that they were about to hear something momentous. When Edward began to speak he accidentally knocked a table leg and the sound was clearly picked up by the microphone, leading millions of listeners to the conclusion that Reith had slammed the door on his way out, his Presbyterian soul black-affronted by Edward's decision to abdicate.

John Reith's autocratic and often bullying leadership made him many political enemies and in 1938 he resigned from the BBC in such a way that his departure did not appear to be entirely of his own volition. He served briefly as an MP and as a wartime Minister and would later hold several company directorships and gain a peerage, but although he remained in the public eye none of his subsequent posts ever matched the influence and impact of his years at the BBC, mainly as nobody was ever willing to give Reith so much power again. While he remained a figure more respected than actually liked he was at least self-aware enough to realise that the Beeb would be his legacy and even after his resignation continued to make noises when he suspected standards he had set might be in danger of falling.

The word 'Reithian' – coined in honour of the Scot who knew nothing about broadcasting before he joined and cared little for television while he was there – remains to this day a word immediately associated with the BBC and his influence can still be seen on public broadcasting around the world. Although one dreads to think what Reith would make of the recent changing of the famous 'inform, educate and entertain' remit to *Snog, Marry, Avoid*.

KEEP THE RED FLAGS FLYING

85 BILL SHANKLY (1914–81)

If Scotland's greatest home victory was the Battle of Bannockburn in 1314, their best away performance is statistically their 5-1 win at Wembley in 1928. This match has gained legendary status and that team, who became known as the Wembley Wizards, are often held up as

Scotland's greatest ever. The two most famous Wizards were Hughie Gallacher and Alex James who both hailed from Bellshill in Lanarkshire. Gallacher was captain of Newcastle United when the Geordies won the English Championship in 1927, and James was the most influential player in the Arsenal team which, in the 1930s, became the dominant and wealthiest team in England and was temporarily renamed the Bank of England club – appropriate when one considers that the Bank of England was founded by a Scotsman.

Arsenal are today one of the most successful clubs in the world; only two English clubs have won more domestic honours. The first, Manchester United, enjoyed their first era at the top when they won the English Championships in 1908 and 1911, with Ayrshire-born Sandy Turnbull their star striker – signed from arch rivals Manchester City – and the first of United's significant Scots. In 1945, after winning no major honours for more than thirty years, United turned once more to a former City player, Matt Busby, who for added frisson had also had a successful spell for Liverpool. Busby, like Gallacher and James, was born in Bellshill, Lanarkshire and at the end of World War II almost joined Liverpool as assistant manger but accepted, instead, the offer to manage United.

Busby's first trophy was the FA Cup in 1948, but he

would not win the League until 1952 thereby ushering in the first of three titles in that decade. The Manchester United team of the '50s were known as the Busby Babes on account of their relative youth and their domination of the game at home – and it was hoped in Europe – was expected to continue for many years. In February 1958, having qualified for the European Cup semi-finals the plane carrying the team crashed on take-off from Munich Airport. Twenty-three passengers died, including eight Manchester United players, and others, Busby among them, were seriously injured.

The 1960s saw Busby's team return to the top of English football with two more Championships in 1967 and the club's most famous forward line: Bobby Charlton, George Best and Denis Law. Law, born in Aberdeen, was another former Manchester City player. He joined the Reds in 1962 and three years later became the only Scot to date to win the European Footballer of the Year award. Law became referred to as The King, but missed, through injury, Busby's greatest achievement when, ten years after the Munich plane disaster, United in 1968 won the European Cup. The following year Busby retired after twenty-four years' service as manager and having established Manchester United as one of the world's most famous and glamorous football clubs.

Within a decade, another English club built up by a

Scotsman became more successful even than Busby's Reds. Bill Shankly hailed from Ayrshire and, in common with Gallacher, James and Busby, he came from a mining background where prowess at football meant a chance to escape a life in the pits. Bill was one of five brothers who played for the local village team, the gloriously named Glenbuck Cherrypickers, and all five went on to play professional football – Bill's older brother, Bob, was manager at Dundee for the team's only Scottish League Championship in 1962.

Bill Shankly chose to head south of the border and he signed for Preston in 1933, but World War II curtailed a promising playing career and in 1949 he moved on to management. He started out with various clubs in the lower leagues and joined Liverpool in 1959, which at that time were also in the lower leagues. Liverpool FC has always featured a strong Scottish presence. When the club was formed in 1892, in a bid to buy success they hired an entire first team squad of Scottish professionals and were known as the Macs; all five of the club's League Championships were won with a Scot as their captain and their greatest player up until then had been forward Billy Liddell, from near Dunfermline. Those glory days had, however, long gone by the time Shankly took over in.

Shankly turned the club around and within three

years Liverpool were promoted to the top division, and between 1964–73 won three League Championships. During his fifteen years as manager, the club introduced the all-red strip which, lifelong socialist as he was, must have pleased Shankly no end. And there was the famous boot room, more than a store for the team's boots, this was where Shankly drank tea with the team's coaches and talked tactics. Liverpool FC and Liverpool supporters were his life – and the devotion he gave was reciprocated by player and fan alike. Of the many quotes attributed and misattributed to the man, the one that sticks in most people's mind is when Shankly, asked if he considered football a matter of life and death, replied, 'It's much more important than that.' And, in an example of his attitude to the game, he told an underperforming player, 'The problem with you son, is all your brains are in your head'. On the subject of his club's arch-rivals he said simply, 'If Everton were playing at the bottom of my garden, I'd pull the curtains.'

In 1974, at the age of sixty, soon after winning the FA Cup, Bill Shankly suddenly announced his retirement. The fans were shocked, but Shankly had put in place a team under the management team of his boot room assistants, Bob Paisley and Joe Fagan, that would go on to become the most successful English football had ever seen. Liverpool would dominate

English football for fifteen years to come and would win the European Cup four times between 1977–84. Subsequent managers gained more honours for Liverpool than had Shankly, but it was he who laid the groundwork and he was the greatest manager the club had ever had.

Bill Shankly lived to see only three of those latter victories; he suddenly died of a heart attack in 1981. He may have come to regret his shock decision in '74 to walk away from the team and the game he loved so much but it did, at least, afford him time with his family and in particular Nessie, his long-suffering wife who, legend has it, was once taken by her husband to watch a Rochdale game as an wedding anniversary present. When someone dared ask Shankly if this could possibly be true he was indignant, 'Of course it is not true. It was her birthday – and it was Rochdale reserves'.

SAND GETS EVERYWHERE

86 DEBORAH KERR (1921–2007)

The genteel town of Helensburgh in Dunbartonshire is famous for, among other things, being the birthplace of John Logie Baird [99], but what is less well known is the town's connection with one of the sexiest moments in cinematic history, the very thought of which would have the residents' lace curtains twitching. Deborah Kerr was born Deborah Kerr-Trimmer and spent only the first few years of her life in Helensburgh before moving with her family to Bristol, where she trained as a ballet dancer before turning to acting. Kerr made her film debut in 1941 and made her name through two films with director Michael Powell, *The Life and Death of Colonel Blimp* in 1943 and *Black Narcissus* in 1947.

Kerr played the lead in *Black Narcissus* and the international success of the film led to her accepting a contract with MGM and moving to Hollywood where she became established as an intelligent, refined and elegant leading lady. Scots would be less pleased that she became depicted as the classic English rather than Scottish rose and that in America her surname was pronounced 'car', apparently to rhyme with 'star', rather than the traditional Scottish pronunciation.

Throughout the 1950s Kerr was one of Hollywood's most successful actresses and almost a guarantee of box-office success for any film she appeared in. She starred in the MGM epic *Quo Vadis* in 1951, *An Affair to Remember* in 1957 with Cary Grant and, also in 1957, played Sister Angela opposite Robert Mitchum in *Heaven Knows, Mr Allison*, but despite, as she was in *Black Narcissus*, being sorely tempted her wimple remained safely intact.

Of all her films, Deborah Kerr is perhaps best known for two – *The King and I*, the Rodgers and Hammerstein musical released in 1956 and the World War II epic, *From Here to Eternity*, released in 1953. For her role as the governess, Anna, opposite Yul Brynner's King Mongkut of Siam in *The King and I*, Kerr won the Golden Globe for best actress for her spirited performance, even though it transpired that her vocal had been dubbed by soprano Marni Nixon.

The Oscar-winning drama, *From Here to Eternity*, was shot in black and white and a departure from Kerr's career so far that had been identified by her reserved persona and her famous red hair. Here, she plays a wife who has an affair with her husband's subordinate, played by Burt Lancaster, and for all that the film features the attack on Pearl Harbour and also stars Montgomery Clift and Frank Sinatra, it is the scene with Lancaster and Kerr, as semi-clothed as it was possible to be in 1953 and kissing in the Hawaiian surf that has become the film's most iconic moment, and has ever since inspired thousands of romantic couples to check out their local tide times.

Deborah Kerr retired from films in 1969 to concentrate on stage and television work. Always the consummate professional, she was nominated six times for the Academy Award for best actress but was never the winner. However in 1994 she finally received an

Honorary Oscar in recognition of a 'motion picture career that has always stood for perfection, discipline and elegance', which led to much happiness for her countless fans and saw the good people of Helensburgh throw away their inhibitions and head to the beach where having had their flask of tea and sandwiches they return swiftly home so as not to miss *Countdown*.

COUTHY GALORE

87 SEAN CONNERY (1930-)

The expression, 'in the limelight', dates back to 1837 when limelight lamps were first used to illuminate the Covent Garden Theatre in London – light was emitted when an oxygen-hydrogen flame was directed at a block of quicklime – and its use caught on quickly in theatres throughout the land. With this, limelight came to be called Drummond light, after Thomas Drummond, a civil engineer from Edinburgh, who although he did not discover it was the first to recognise that the light's intense brightness would be ideal for use during the surveys he conducted as an engineer and had, in 1826, built the first working model of his lamp.

The name Drummond light fell out of fashion but in the 21st-century cinematic limelight there have been an unprecedented number of great Scottish actors. Brian

Cox from Dundee, Gerard Butler from Paisley and James McAvoy from Port Glasgow are but three of a plethora of internationally acclaimed Scottish stars, and in this age of sequels, prequels and reboots three movie franchises stand, like the Hollywood sign, high above the rest and all three boast a Scottish connection. Robbie Coltrane from Rutherglen and the Glenfinnan Viaduct both feature prominently in the Harry Potter films and Ewan McGregor from Crieff plays the young Obi-Wan Kenobi in the underwhelming, but still disconcertingly lucrative *Star Wars* prequel trilogy. The prequels may not be the best loved of the *Star Wars* canon, but the Emperor was played by Ian McDiarmid from Carnoustie in Angus, which gave us the opportunity to see a McGregor against a McDiarmid, just as it was a long time ago on a Scottish battlefield far, far away.

The third of these all-time great movie franchises began in 1962 and is still going strong nearly fifty years later. In the twenty-two official releases in the series to date, six actors have played the leading role but as far as most people are concerned there will only ever be one James Bond, and he comes from Edinburgh. And if the career of Thomas Sean Connery teaches his fellow Scottish actors anything, when it comes to attempting accents never say never without rolling your Rrrrs.

Sean Connery has been a Hollywood leading man

for four decades: he starred for director Alfred Hitchcock in *Marnie* in 1964 and for John Huston in *The Man Who Would Be King* in 1975; he starred as a Franciscan monk in *The Name of the Rose* in 1986, as a Lithuanian submarine commander in *The Hunt for Red October* in 1990 and as Harrison Ford's irascible Dad in *Indiana Jones and the Last Crusade* in 1989. In his long and illustrious film career Connery rarely had the opportunity to appear in a Scottish film, but did have a supporting role in the cult classic, *Highlander* series from 1986; in one of the more incongruous pieces of Hollywood casting Connery played a 2,000-year-old Spanish-Egyptian nobleman, while the immortal Highlander was played by a Frenchman (Christopher Lambert), but then considering that said Highlander can only be killed by decapitation then having a Spaniard with an Edinburgh accent backpacking around Glencoe was not actually too much of a stretch. At the 1988 Academy Awards Connery won the Oscar for best supporting actor for the previous year's *The Untouchables*, but the award was as much for his remarkable career and in particular for his seven appearances as 007 that remain the highlight of his career.

Ian Fleming's first novel to feature the James Bond character was *Casino Royale*, written in 1953, and this and his subsequent novels on the same theme formed the basis for the majority of the Bond film titles. Fleming was born in London but was descended from the Scottish Fleming family of merchant bankers. Fleming served in World War II as a Naval intelligence officer in charge of various top-secret commando units and drew on the experience to create his British secret agent and spy, James Bond.

The jury is out regarding which, if any, of Fleming's contemporaries might have been the inspiration for Bond and many, many names have been put forward, including commando leader Simon Fraser, Lord Lovat, from Inverness, who led the 1st Special Service Brigade during the Normandy landings and Simon Fraser's cousin, David Stirling from Bridge of Allan; in 1941 Stirling became the first commander of a unit operating exclusively behind enemy lines, a new commando unit named the Special Air Service, or SAS. But Fleming revealed no name and, for more people, Sean Connery *is* James Bond. Indeed, the author demonstrated his approval in his 1963 novel *On Her Majesty's Secret Service* when he gave Bond a Scottish father, although tantalisingly never stated a birthplace for wee Jimmy.

Sean Connery was born and brought up in Edinburgh and, before embarking on his acting career, had worked as a milkman, a lorry driver and enlisted in the Royal

Navy; he was, also, offered a trial by a professional football team and had entered various bodybuilding competitions. When Connery was offered the lead in the first Bond film, *Dr No*, as an actor he was a relative unknown and many doubted that he was ready for the role. However Connery's charismatic performance proved key to the film's global success and *Dr No* was followed by the even more popular *From Russia With Love* in 1963, *Goldfinger* in 1964 and *Thunderball* in 1965. Within the Bond genre, this quartet remains unsurpassed for memorable villains, iconic leading ladies and sardonic humour but, above all, it was Connery's Bond who captured the public's imagination.

Sean Connery's penultimate portrayal of the character he made his own was in *Diamonds Are Forever*, released in 1971. When *On Her Majesty's Secret Service* was released in 1969 it was actor George Lazenby's Bond who ended up getting married only to have his wife killed just a few minutes later which, Lazenby pointed out wearily, 'never happened to the other feller'.

TOP GIGHA

88 JAMES (JIM) CLARK (1936–68)

Scotland's love affair with the motorcar, and fast cars in particular, has been long and lasting; David Dunbar

Buick [59] from Arbroath, for example, was the Buick behind the iconic American marque. Much more recently, the video games' market capitalised on the Scots' passion for speed behind the wheel when, in 1997, Dundee-based DMA Designs (now Rockstar North from Edinburgh) launched *Grand Theft Auto*, which – seventy million copies later – is that market's eighth biggest seller of all time and has helped place Scotland at the forefront of the global games industry – even if the makers decided to set the franchise in America rather their hometown as they realised that there were no cars in Dundee worth stealing.

The Scots' love for acceleration is best illustrated by the country's very considerable success in motor sport: Dario Franchitti from Bathgate in West Lothian is a household name in the US, having twice won the IndyCar series as well as twice winning the Indianapolis 500 – America's premier race – in 2007 and 2010; the late Colin McRae from Lanark won the World Rally Championship in 1995 and David Coulthard from Dumfries and Galloway managed, somehow, in his fifteen-year Formula 1 career to win thirteen Grand Prix titles, with most of the rest of his time spent trying to avoid colliding with Michael Schumacher.

And among the all-time greats we have, of course, Jackie Stewart from Dunbartonshire. Between 1969–73

Stewart won the F1 World Championship three times and when he retired after his third triumph in 1973 his total of twenty-seven Grand Prix victories was a world record. He was known as the Flying Scot and after nearly dying in a crash in Belgium was ahead of his time in being a tireless campaigner for improved safety on race circuits. Statistically, Jackie Stewart remains the most successful British driver in the history of F1 but when it comes to selecting the greatest racing driver of them all Stewart is overtaken by the claims of others, including Argentina's Juan Manuel Fangio, Germany's Michael Schumacher and two others who during their short careers were seen as the most effortlessly talented and gifted drivers the sport has ever seen: Ayrton Senna from Brazil and a sheep farmer from the Scottish Borders by the name of Jim Clark.

James – much better known as Jim or Jimmy – Clark was born in Fife, but moved with his family to a farm in Berwickshire. Clark learned to drive at an early age and in the mid-1950s began racing his own car in local rallies. He had four sisters and his parents were less than keen on the amount of time their only son was spending away from work on the farm. Their son's talent for driving was, however, undeniable; he soon progressed up through the various competitive race classes and, in 1960, began his Formula 1 (F1) career with Lotus for whom he drove for the remainder of his career.

In 1961 Clark was involved in one of the worst accidents in F1 history and was so deeply affected by it that he seriously considered giving up motorsport. At the Italian Grand Prix at Monza, a collision between Clark and Wolfgang von Trips propelled the German driver's car into the crowd killing von Trips and fourteen spectators. But Jim Clark eventually decided to continue in motor sport and the following year he won his first Grand Prix and came second overall in the world Championship. Over the next years Clark became the undisputed top driver in F1, winning the world Championship in both 1963 and '65.

But throughout his F1 career Clark was bedevilled by gear and engine problems, his record number of pole positions were testament to his ability to be the fastest but the Lotus was unreliable. Nevertheless, he remained loyal to the team that gave him his first F1 opportunity and, overall, won twenty-five Grand Prix. He won international fame and respect for his consummate skill as a driver and remains, to this day, ranked among the all-time giants of the sport.

Furthermore, in Clark's day F1 drivers also competed in races outside the grand prix circuit: in 1965 he won the Indianapolis 500, marking the only time the Indy 500 and the F1 World Championship have been won by the same driver in the same year. However in April 1968, during a Formula 2 race at Hockenheimn in Germany Clark's Lotus skidded off the track and crashed into trees. Jim Clark sustained horrendous injuries and died before reaching hospital.

At heart a quiet, shy man, Clark was ill at ease with the glitz and glamour associated with motor racing and he felt embarrassed by the attention his fame brought. His funeral took place at his home village, Chirnside in the Borders, at the church next to the farm where he was brought up and to which, even at the height of his fame, he had regularly returned. Jim Clark may have been one of the greatest and most natural racing drivers the world has ever seen, but when it all comes down to it, he was a farmer from a farming family who just happened to be very good at driving fast.

THEY MAKE PROPER WATCHES IN GOVAN

89 ALEX FERGUSON (1941–)

As every Scottish football fan knows, the first British team to win the European Cup was Celtic when they defeated the favourites, Inter Milan, 2-1 in Lisbon in 1967. All of the Lisbon Lions were born within thirty miles of Glasgow, and the radius would have been remarkably even smaller if No. 11, Bobby Lennox, did not have good fortune to hail from Saltcoats in Ayrshire. The Celtic manager was Jock Stein from Lanarkshire who, alongside Matt Busby and Bill Shankly [85], was one of the legendary triumvirate of Scottish managers who dominated British football from the 1950s through to the 70s. And when English clubs were at the fore in Europe between 1977–84, much of their success was down to Scots: John McGovern from Montrose captained Nottingham Forest to their two European Cup victories in '79 and '80, and in those halcyon days every major club in England had at least three Scottish players in its team, all of whom had perms. Liverpool won the European Cup four times, with, in 1984, the team captained by Graeme Souness from Edinburgh and in three of those victories Liverpool's greatest ever player, Kenny Dalglish from Glasgow, was in the team.

Dalglish is the only player to have won more than a hundred caps for Scotland. He joined Liverpool from Celtic in 1977 as the replacement for Kevin Keegan, who had transferred to Hamburg. Keegan was England captain and Europe's top player, but Dalglish would far surpass him by winning six League Championships and three European Cups as a player, plus another three Championships as manager. He is generally recognised as Scotland's greatest ever player, even if the man himself would be reluctant to venture more than a, 'Mebbes aye, mebbes naw'. When Daglish was manager of Liverpool and then Blackburn Rovers between 1985–95 he was joined at the very top of English football by a resurgent Arsenal who, managed by George Graham from Lanarkshire, won two League Championships, but it was another man from the west of Scotland who would, as a manager, surpass them all – Alex Ferguson, born and raised in Govan.

As a young man, Ferguson initially combined an amateur football career with work as a tool-maker in the shipyards where he became a shop steward and would be seen regularly looking at his watch to make sure that all overtime was properly recorded. Eventually, Ferguson turned professional and became an uncompromising striker for St Johnstone and Dunfermline, where his elbows were as sharp as his football brain, but a big money move to Rangers in 1966 was to prove a disappointment for everyone except the Celtic captain Billy McNeill, who Ferguson was supposed to be marking when McNeill scored the opening goal in the 1969 Scottish Cup Final. Chastened, but unbowed Ferguson left Rangers and continued his career as a player until 1974 when he became manager of the far from mighty East Stirlingshire FC, or The Shire as they are known by their dozens of supporters.

The manager soon moved on to St Mirren and then, in 1978, to Aberdeen. He led the Dandy Dons into the most successful period in their history by winning three Scottish Championships and temporarily interrupting the Old Firm's divine right to rule. In the wake of Jock Stein's untimely death, Ferguson was Scotland manager at the 1986 World Cup Finals in Mexico, but it was Aberdeen's victory in the 1983 European Cup Winners' Cup that truly established his international reputation. The final was played in Gothenburg and, against all expectations Aberdeen defeated the mighty Real Madrid. For a glorious few days the Swedish city relived the time when, centuries before, Gothenburg was one of Scotland's major trading ports and home to hundreds of its countrymen. Although the Aberdonians were shocked to find that alcohol was no longer sold at 18th-century prices.

Alex Ferguson was appointed Manchester United manager in 1986. The club had not won the English Championship or European Cup since Matt Busby retired in 1969 and Ferguson's early years with the club proved difficult. By 1990 with no trophies and no sign of progress it seemed probable that Old Trafford would be saying 'ta ra' to Fergie. However, there are lucky managers and then there is George Burley, and United would somehow win the FA Cup in 1990 and three years later their first League Championship in twenty-six years. They went on to win eleven English Premier

Leagues (the successor to the English League Championship) between 1993–2009, six more than Matt Busby had achieved with the team.

During those years, Ferguson would build at least three great teams through inspired signings such as Peter Schmeichel, Eric Cantona, Roy Keane, Cristiano Ronaldo and Wayne Rooney. And, following the tradition he had begun at Aberdeen of bringing through talented young players, he oversaw the careers of Ryan Giggs, David Beckham and Paul Scholes. Ferguson's apparently impossible aim when he took over at United was to replace Liverpool as England's top team. This, of course, he achieved and went on to turn Manchester United into one of the biggest, if not the biggest, football clubs in the world.

In 2001, having won the English Premier League for the third year in succession, Alex Ferguson announced that he would retire in 2002 to spend more time with his family. Then he thought about it, and realising there were only so many letters of complaint one man could write to the BBC, he swiftly changed his mind and in 2008 in Moscow became the only Scot to manage two winning Champions League teams – although wasn't it a shame when that nice John Terry missed his penalty.

At the close of 2010 Ferguson will equal Matt Busby's twenty-four years as manager of Manchester United and has become the undisputed Godfather of football, revered and adored by United's millions of fans worldwide and considered an unholy combination of Beelzebub, the Antichrist and DCI Taggart by everybody else. It is unclear how many more years, or decades, Ferguson will continue to rule at Old Trafford, but, for now, United's team of Brazilians, Bulgarians, Ecuadorians and Koreans battle to uphold the exacting standards expected by their manager – even if they have no idea what he's saying to them – and the loyal and devoted fans of East Stirlingshire FC where football truly is bloody hell can only pause and ponder what might have been.

FOR THOSE ABOUT TO WEAR SHORT TROUSERS (WE SALUTE YOU)

90 ANGUS YOUNG (1955–)

The English are a remarkably tolerant nation when it comes to their Scottish neighbours. They continue to bank with Scottish banks even when said Scottish banks have nearly bankrupted the country and they continue to instruct their commentators to bend over backward to call English sportsmen British and Scottish sportsmen Scottish despite getting absolutely no thanks from Scots for doing so. Another example of this largely unrecorded benevolence can be found in the 1960s when the world rocked and rolled to the sound of what would become known as the British Invasion – even though almost all the bands came from Liverpool, London, Newcastle, Manchester, Birmingham and every other English town and city you can think of. However, if they were peripheral at best in the British Invasion of the 1960s, when the second British Invasion came around in the '80s Scots were ready and willing to wear as much eyeliner, hair spray and oversize jackets with strategically rolled up sleeves as necessary to gain a slice of the MTV action.

It was Sheena Easton from Bellshill who was the first to make an impact by reaching No. 1 in the US in 1981 with *Morning Train (Nine To Five)*. Easton would have further success by singing a James Bond theme tune, collaborating with Squiggle and then relocating permanently to the United States, a decision that appeared to please her American and British fans equally well. Two years later, Annie Lennox from Aberdeen would also reach No. 1 in the US with *Sweet Dreams (Are Made of This)* as the interesting one of the duo the Eurythmics. Lennox became an icon through her powerful vocals, groundbreaking videos and even more groundbreaking androgynous visual style that culminated in a performance at the Grammy Awards dressed as

Elvis – to the utter astonishment of many American viewers who could have sworn they had seen The King at their local supermarket just the day before. Lennox enjoyed success in the Eurythmics and as a solo artist and her Aberdonian heritage was reflected in songs such as *Here Comes the Rain Again* and *You Have Placed A Chill In My Heart*.

In 1985 Glasgow band Simple Minds who had already achieved considerable success and critical acclaim in Britain reached No. 1 in the US and around the world with *Don't You (Forget About Me)* from the film *The Breakfast Club*. The single's unexpected success would propel Simple Minds to international stardom and the song became one of the defining records of the decade, which was all very well at the time, but not much use when 1990 came along and despite the title of the song collective amnesia set in.

If Sheena Easton, Annie Lennox and Simple Minds were clearly categorised as Scottish artists, other Scots from slightly further afield also gained international success in the 1980s. In 1983 Men At Work from Melbourne achieved the very rare feat of being No. 1 in both the singles and albums charts of the UK and US with *Down Under* and *Business as Usual* respectively. *Down Under* became an Aussie anthem; complete with references to vegemite sandwiches, even though the band's lead singer and main songwriter was Colin Hay from Kilwinning in Ayrshire, a place not known for vegemite sandwiches, or for that matter, vegetables.

Dire Straits, who had huge success with the singles *Sultans of Swing* and *Money For Nothing*, and their 1985 album *Brothers in Arms* which sold more than twenty million copies worldwide, are mostly associated with Newcastle upon Tyne, where the band's co-founders, Mark and David Knopfler, grew up; however both the Knopfler brothers were born in Glasgow and moved south at an early age. It has been estimated that Dire Straits have sold an incredible one hundred million records, even if nobody has actually listened to one since around 1992. Despite this, the Knopfler brothers are not the most successful Scottish recording artists of all time.

George, Malcolm and Angus Young were all born in Glasgow before immigrating to Sydney, Australia as a family in 1963. George, the eldest, was the first to find success when he joined the Easybeats as the band's rhythm guitarist and they had an international hit in 1966 with *Friday On My Mind*. However, when his kid brothers formed the rock band AC/DC in 1973, with Angus on lead guitar and Malcolm on rhythm, George had to make do with production and management duties. In 1974 the band recruited Bon Scott, who was born in Kirriemuir in Angus but grew up in Western Australia, as their lead singer and released their first album, the appropriately titled *High Voltage* the following year.

AC/DC made their international breakthrough with *Highway to Hell* in 1979 and by this stage had established their reputation for hard rock anthems, powerful riffs so catchy that even the punks were impressed, and classic songs such as *Touch Too Much*. In 1978 they released their famous live album, *If You Want Blood You've Got It*, which was recorded at the Glasgow Apollo with the introductory guitar riff of *Whole Lotta Rosie* interspersed with the crowd chanting, 'Angus, Angus', just to make the lead guitarist feel at home, although a later attempt at chanting 'Malcolm, Malcolm' would sadly prove less successful.

With *Highway To Hell* bringing the band their first chart success in the US and mega-stardom awaiting them, in February 1980 Bon Scott at the age of only thirty-three died in London of acute alcohol poisoning. His shocked band-mates briefly considered quitting, but the Young brothers decided to continue and with English singer Brian Johnson taking on vocal duties they recorded their biggest selling album to date, *Back in Black*, which was released in the same year. *Back In Black* has sold more than twenty million copies in the US alone and become one of the five best-selling albums of all time worldwide. More No. 1 albums followed and the band continued to stage mammoth, international, sell-out tours over the next thirty years. It has been estimated that AC/DC have sold more than 150 million records worldwide and in the process have inspired generations of teenagers to play air guitar in their bedrooms and dream of the moment when they, too, will be shaken all night long.

Throughout the band's long career it is the Young brothers, Angus and Malcolm, who have remained the AC/DC constants. While older brother Malcolm has remained an understated performer on stage, Angus Young has been the hyperactive focal point. Never the tallest of men, when AC/DC were formed Angus decided to make a virtue of this fact by dressing up on stage in the uniform of his former school, complete with short trousers, tie and schoolboy cap. AC/DC have now been playing for more than thirty-five years and despite Angus Young now being in his mid-fifties he continues to wear his trademark schoolboy outfit and, to the delight of millions of fans, he still strips down to his underwear at some stage during concerts. Hell's bells indeed.

Innovative Scots

So, we come to the final category and the final ten Scots to make my top one hundred. Innovative Scots are those who have, through politics, or conflict, or communications, or science or invention, either brought the world closer together or pushed it slightly farther apart. Or to put it another way, these are the ten Scots who didn't fit in to any of the other categories.

STAIRWAY TO IRVINE

91 JOHN PAUL JONES (1747–92)

For a country with such a long seafaring tradition, the history of the Scottish navy is surprisingly meagre.

Comparable countries such as the Netherlands, Portugal and Sweden built up navies which were the equal of any of the larger European powers, they circumnavigated the globe and founded colonies in South America, Africa and Asia. When the Royal Scots Navy merged with the English Royal Navy in 1707 there were nearly 300 ships in the English fleet and only three in the Scottish; not so much a Scottish Armada, more a modest boating club.

For ambitious Scottish sailors who wanted more from a life at sea than salted herring there was little option but to look elsewhere. Samuel Greig from Inverkeithing in Fife joined the Imperial Russian Navy of Catherine the Great in 1763 as a captain, and by the time of his death in 1788 had been promoted to admiral and was accorded a state funeral in recognition of his achievement in turning the Russian Navy into the most powerful in Eastern Europe. Another Scot who gained a different form of recognition, was William Kidd from Greenock. Kidd was a successful trader and ship owner based in colonial New York who in 1695 was given a commission by the English Government to deal with piracy in the Indian Ocean. Captain Kidd adopted a novel strategy of attacking and plundering the merchant ships himself, which was not quite what the authorities had in mind. Despite pleading his innocence, Kidd was hanged for piracy in 1701, although his treasure has never been found. Har-har.

At least Kidd was given credit for being a competent seaman. Alexander Selkirk from Largo in Fife, however, was not, and when sailing in the Pacific so infuriated his captain that he was put ashore on the uninhabited island of Juan Fernandez, 400 miles (644 km) off the coast of Chile, in 1704. Selkirk remained on the island, surviving on a plentiful supply of wild goats, until he was finally rescued in 1709, by which time his goat's cheese and red onion tartlets were quite sublime. The story of the castaway Selkirk inspired Daniel Defoe's 1721 novel, *Robinson Crusoe*, which, in turn, inspired numerous books, films and television series as well as the long-running BBC radio programme *Desert Island Discs*. Selkirk did have a Bible with him when he was put ashore on the island from which he gained great strength and support, but would soon regret the selection of Scottish records that he took with him, as there is only so much of the Proclaimers that even the proudest of Scotsmen can endure.

John Paul Jones was born in Kirkcudbright, his birth-name was simply John Paul and he went to sea when he was 12 years old. He sailed on slave ships and progressed to the rank of captain before he was forced to flee to America after running a man through with his sword, following a dispute over wages. He was, therefore, clearly an ideal candidate to join, in 1775, the newly formed revolutionary Continental Navy, the precursor to the United States Navy, and was given the rank of first lieutenant. It was, also, around this time that John Paul changed his name to Paul Jones, as he had become fed up with having to kiss the tarmac every time he set foot in a new country.

Jones's first American ship was the *USS Alfred* and he was given the honour of being the first to hoist the new Continental flag in December 1775 the precursor of the Stars and Stripes. Jones was soon promoted to captain and then commodore and, aboard the *USS Ranger*, was sent to France, where after the French agreed to support the American rebels in the War of Independence against the British, he spent the next two years raiding the British coastline and captured both *HMS Drake* and later *HMS Serapis*, the first American successes in British waters.

John Paul Jones became a hero of the American Revolution and his naval exploits gained him the title, Father of the American Navy. Yet to the British he was no more than a pirate, an epithet reinforced by his decision when raiding his Kirkcudbright homeland, to go ashore and kidnap the Earl of Selkirk, although perhaps fortuitously the Earl was out for dinner that evening. A biographical film about Jones's life was made in 1959 and his exploits would inspire a young English bass guitar and keyboard player by the name of John Baldwin to change his name to John Paul Jones. The musician, in turn, would find fame as a member of the band Led Zeppelin – who were also not averse to creating a little mayhem in their time.

FOR AS LONG AS A HUNDRED OF US REMAIN ALIVE . . . WE WILL ALWAYS REQUIRE A DAILY POSTAL SERVICE

92 JAMES CHALMERS (1782–1853)

The town of Arbroath in Angus has several claims to fame including its barrel-smoked haddock, known as the Arbroath Smokie. The town is also famous for the Declaration of Arbroath, the most important document in Scottish history, which was drafted by Bernard de Linton, the Abbot of Arbroath Abbey, in 1320.

Not so well known is that Arbroath was the birthplace of James Chalmers, who was the inventor of the adhesive postage stamp. Chalmers moved to Dundee where he made his name as a bookseller, printer and local councillor and for years campaigned for reform of the British postal service. Payment for mail at the time was the responsibility of the recipient who was liable to various regional rates and the distance that the letter had travelled. This made receiving a letter an expensive business, especially if it had crossed the Scotland-England border. And if the recipient suspected that the postman was the bearer of bad news, had taken a longer than necessary route or had designs on his wife or daughter, then the recipient had the right to refuse the delivery.

Chalmers worked on ways to improve this system and in 1834 at his Dundee printing works came up with samples of adhesive labels, or stamps, which could be stuck on a letter and indicated that one-penny had been pre-paid. Unfortunately for posterity Chalmers did not publish any details about the world's first postage stamp, but continued to write to the relevant authorities proposing a universal, pre-paid postage system with payment based on weight rather than distance – although the authorities remained in a position to refuse delivery of these letters if they recognised they came from that Dundee bloke banging on about the postal service again.

It was Englishman Rowland Hill who, independently of Chalmers, published a pamphlet in 1837 calling for universal pre-paid postage and it was Hill, and not Chalmers, who gained the credit in 1840 for the world's first nationally approved pre-paid postage stamp. The stamp featured an image of Queen Victoria on a black background, as chosen by Hill, and it became known as the Penny Black. The name of James Chalmers was all but forgotten when discussing the origins of the postage stamp, but in recent years there have been belated calls that Chalmers, too, should be honoured for his contribution by appearing on a stamp, but sadly the correspondence appears to have got lost in the post.

HOLY INK CARTRIDGES, FAXMAN!

93 ALEXANDER BAIN (1811–77)

For readers under the age of twenty-five who may have wondered what that strange machine is that sits all alone in the corner of the office and whose only purpose seems to be to make intermittent screeching noises and chew paper, well, this is what your older colleagues call a fax machine. Hard as it may be to believe, in the 1980s and '90s the fax machine, or the facsimile machine to give it its full title, was at the cutting edge of communications. No up-and-coming business could be said to have truly made it without owning said machine, though many finance directors were somewhat reluctant to embrace the prospect of the faxed invoice – which arrived so much sooner than those sent by post – as the next thing you knew was that suppliers would be expecting to be paid on time.

The fax machine became popular in the 1960s and '70s as a result of developments in the US and Japan which enabled faxes to be sent along telephone lines and the machines becoming more compact and affordable. Yet the technology on which the fax machine is based dates back to the 1840s, thirty years before Alexander Graham Bell [96] began his research on whether people wanted new kitchens, windows or doors as representatives just happened to be in the area – and, all those years ago, it was a Scotsman who discovered that it was good to fax.

Alexander Bain was born in Caithness, the son of a crofter, but was fascinated not by sheep but by clocks. However, as time seems to tick by very slowly in Caithness he moved to England to train as a watchmaker, and in 1837 began working in London with Charles Wheatstone, one of the co-founders of the world's first commercial telegraph company. Bain was intrigued by the newly developing technologies of electricity and the telegraph and determined to advance the world of communications. In 1841 he patented the world's first electric clock, despite his employers attempts to claim his invention as their own which had resulted in a lengthy court case. In place of the traditional springs and weights, Bain's clock had an electromagnetic pendulum to keep it running and was the forerunner of all modern electrically driven and battery-run timepieces.

Bain, then, set about improving the electrical telegraph that his erstwhile former employer Charles Wheatstone had co-invented in the UK in 1837, with American Samuel Morse independently inventing his own electrical telegraph in the same year. His first innovation was to put his electric clocks at either end of the telegraph wire, which meant that the time could be synchronised countrywide: thus, five o' clock in Glasgow would be five o' clock in Edinburgh and five o' clock in Aberdeen, although it would always remain half past two in Caithness. Emboldened Bain also became convinced that it would be possible not only to send messages by telegraph, but that images could be sent too. To achieve this, he had two pendulums at each end of the telegraph wire with a stylus on each, one sending and one receiving, and these he connected by wires. The sending pendulum swung across raised metal type and every time its stylus touched the type, it sent an electric current along the wire to the receiving pendulum at the other end, which would swing across chemically soaked paper on which its stylus would leave a mark.

In 1843 Bain received a patent for '. . . improvements in timepieces and in electric printing and signal telegraphs', in other words, the world's first fax machine. When one of his chemical telegraphs was demonstrated in France, sending messages between Lille and Paris, it proved five times faster than any existing telegraph system. Bain worked on, and in 1846 developed a mechanism which would punch transmitted signals onto paper tape, an innovation which would speed up telegraph transmission immeasurably and was the forerunner of the ticker-tape machine – so called for the ticking sound it made.

Throughout the 1840s he patented numerous inventions and the money rolled in, in no small part due to his readiness to take legal action, as he had against Charles Wheatstone, to protect ownership. Bain was, however, an uncompromising man and people found him difficult to do business with. He lost money as quickly as he made it and by the 1850s Alexander Bain was bankrupt and his work on telegraphic systems was rejected in favour of that of others. After moving to America and finding little success, he returned to Scotland – where there was no ticker-tape reception – and spent the remainder of his life an impoverished and disillusioned man waiting sadly by his most famous invention for the fax that never came.

NO SLEEP TILL BRIDGETON

94 ALLAN PINKERTON (1819–84)

The most famous policemen in Scotland surely have to be Detective Chief Inspector Jim Taggart in the eponymously named television series set in Glasgow, and Inspector John Rebus in the Edinburgh-based crime novels by Fife-born author Ian Rankin. The most influential Scottish detective, however, does not reside in fiction. In the classic Western, *Butch Cassidy and the Sundance Kid*, when Butch and the Kid, having held up a train, are being tracked doggedly, day in and day out, by a posse of horsemen Butch (Paul Newman) wants to know, 'Who are those guys?'. Well, the answer is that they are Pinkerton agents who take their name from a Glaswegian cooper, rather than a Glaswegian copper, called Allan Pinkerton.

As well as being a cooper Pinkerton was a political radical in his hometown and when his activities looked as if they would result in imminent imprisonment he emigrated. The year was 1842, he was twenty-three years old and he headed for Canada and then the US. Pinkerton began his sleuthing career in Chicago, where in 1850 he founded what would become the Pinkerton National Detective Agency and initiated methods such as surveillance and going undercover which, of course, have become standard practice. Pinkerton gained repute in 1861 when he was employed to deal with security for President-elect Abraham Lincoln and suspecting an assassination attempt had been planned in Baltimore persuaded Lincoln, who was travelling to his inauguration, to go non-stop through the town and at night. History debates whether or not the threat was genuine and Lincoln was much criticised for his perceived cowardice, but Pinkerton had just watched the first series of *The Wire* and was taking no chances.

After the Civil War Pinkerton saw his agency grow to become the largest detective and security firm in the United States. One of its specialities was the pursuit of train robbers and outlaws for which it was renowned and, more controversially considering Pinkerton's earlier political radicalism in Scotland, the agency's men were employed by businessmen to break up labour disputes by force. This violence in the late 19th century against the early American unions employed by the agency men would cause long-standing damage to the Pinkerton name.

Thankfully Allan Pinkerton was by then no longer around to see his agency that he had built up from nothing so tarnished. In the latter part of his life he concentrated on promoting his company and wrote a series of books celebrating his work as a detective.

Pinkerton agents became known as Private Eyes – although many would call them much worse – and although the agency would decline in the 20th century, the concept of the private eye continued to grow through films, books, television series and a certain British satirical magazine.

The name private eye originated from the agency's famous logo depicting an eye and the slogan, We Never Sleep, and emblazoned across the Chicago headquarters. The logo was devised by Allan Pinkerton himself and many concerned Chicago citizens to write in to the agency with suggestions of an evening bath or a glass of hot milk in order to help the detectives with their insomnia.

THE TIME IT IS A-CHANGIN'

95 SANDFORD FLEMING (1827–1915)

There is an old British Rail strapline, 'Let the train take the strain', and in 19th-century North America the strain of creating the rail line crossing the continent from east to west was a job that seemed ideally suited to Scots. In 1862, in the midst of the American Civil War, Daniel Craig McCallum, an engineer and bridge-builder who had emigrated from Johnstone in Renfrewshire, was appointed Military Director and Superintendent of

Railroads for the entire railway and telegraph line network controlled by the Union forces – the largest railway system in the world at the time. McCallum remained in position until the end of the war in 1865 and the Union's ultimate victory was due in no small part to the fact that he managed to keep the rail lines open for business no matter how many Confederate leaves happened to be on the line.

The vision, ingenuity and perseverance of Scots in the railway industry is impressive to say the least – one day perhaps it might even be possible to build a railway from Glasgow to Glasgow Airport or a tram-line from one end of Edinburgh Princes Street to the other. Over and above McCallum's contribution, the Canadian Pacific Railway – opened in 1885 – was commissioned by two Scottish-born Canadian prime ministers, financed by a consortium of Scottish-born businessmen and its chief engineer, Sandford Fleming, was born in Kirkcaldy.

Sandford Fleming emigrated to Canada in 1846 at the age of eighteen and found work as an engineer, surveyor and businessman, and he would spend many years, at first advocating and subsequently working on a railway line that would link the continent from west to east. His interests and repute, however, extended far beyond Canada's rail network and included, for example, in 1851 designing Canada's first ever postage stamp –

the three-penny beaver. But it is for his work on Standard Universal Time that he is best known. It would perhaps be pushing it a little to claim time itself as a Scottish invention, but before Fleming came along the question, 'What time is it?', was almost impossible to answer.

The time was whatever the time was at wherever you happened to be, or to put it another way – local time for local people. Anyone travelling across North America, for example, would need to re-set their watch as they moved from one local time zone to the next – and there were more than a few. In addition, time was expressed in hours but only from one to twelve with a.m. and p.m. to define before and after midday. In 1876 Fleming visited Europe and, one day, happened to be travelling by train in the north of Ireland. He arrived in plenty of time for the 5:35 p.m. train from the Donegal seaside town of Bundoran, but when the train didn't come and he rechecked his timetable, he realised it was scheduled for 5:35 in the morning. He put his mind to solving such problems for the future and, three years later, proposed a radical but inherently simple twofold solution: first, every country in the world should adopt a twenty-four-hour clock, and secondly, the globe should be divided into twenty-four time zones, each one hour apart and also 15° longitude apart – which adds up to 360°. In 1882 at an international conference Fleming's proposal for a universal 24-hour clock was accepted, and to the chagrin of the French Greenwich in London was selected as the centre of the world in 1883; the meridian, 0°.

It would take until 1929 before international time zones were universally accepted and political considerations have resulted in deviations from the original plan: Russia has a rather greedy eleven time zones, while both India and China seem more than happy with one zone each and despite many decades of improved diplomatic relations Spain and Portugal still cannot bear to be in the same time zone as each other.

MR BELL CAN'T GET TO THE PHONE RIGHT NOW, PLEASE LEAVE YOUR LEGAL CHALLENGE AFTER THE TONE

96 ALEXANDER GRAHAM BELL (1847–1922)

If Alexander Bain [93] was the forgotten genius so ahead of his time that it would be more than a century before the invention he pioneered, the fax machine, was fully appreciated, then Alexander Graham Bell was the Scot in the right place at the right time. Bell's telephone had a huge and immediate impact which made him, along with James Watt [52] and John Logie Baird [99] not only one of the best known of the Scottish inventors, but also one of the best known inventors in the world.

Bell's career exemplified Scottish ingenuity and innovation, alongside the Scots' uncanny ability in the face of competition to be the first to patent, publicise and demonstrate, even when evidence suggests that said Scot may not actually be the inventor – as it did in Bell's case.

Alexander Graham Bell was born in Edinburgh where his father was a teacher of the deaf and his mother was also deaf so it is perhaps not too surprising that Alexander's early career mimicked that of his father. The Bells moved to work in London and then Canada in 1870, before Alexander immigrated to the US the following year and in 1873 became Professor of Vocal Physiology at Boston University. He became increasingly obsessed with sound and with the transmission of speech along electric wires and, undeterred by his lack of formal scientific training, began to experiment.

Bell also followed in his father's footsteps by falling in love with one of his deaf pupils, Martha Hubbard, whose father just happened to be a) wealthy, b) willing to fund Bell's experiments and c) keen to find a gadget that would allow him to keep in touch with his aged mama – Old Mother Hubbard. By the end of 1875 Bell was under pressure; the Hubbards were becoming impatient and other inventors, including acclaimed American scientist Elisha Gray, were close to producing working telephones. Bell had managed to transmit some form of sound waves and musical notes down a wire, but it was indistinct and nothing that at that point could be said to resemble speech. Not only Bell's financial status, but also his future marital status were in the balance when, appropriately, on St Valentine's Day in 1876, the grand final of the World Telephone Patenting Competition saw both Bell and Elisha Gray present their patents to the Boston Patent Office.

What happened on 14 February 1876 has remained shrouded in mystery and controversy. We do not know for certain which of Bell or Gray was the first to patent, we do not know for certain what were the grounds that Bell was awarded the patent over Gray and we do not know when and how a diagram of a liquid transmitter that was in Gray's original patent happened to appear almost identically similar in a later diagram by Bell. Suspected industrial espionage notwithstanding, Bell was awarded the patent and on 10 March 1876, Bell gave the first public demonstration of his telephone – that included a liquid transmitter – and down the line to his assistant, Thomas Watson, who was in another room in the building, uttered the famous words, 'Mr Watson, come here, I want to see you', much to the disgruntlement of Watson who had been up and down the stairs all day.

In 1877 the Bell Telephone Company was established

and the inventor finally got around to marrying Martha who, atypically (through sign language of course), was the one complaining that her husband was always on the phone. Unsurprisingly, there were numerous legal challenges to Bell's patent, but in every case was upheld in his favour. By the end of the 1880s more than 150,000 Americans had bought telephones from the Bell Telephone Company and the rest of the world would soon follow suit from their phone provider of choice. 1915 saw the advent of trans-continental calls and the historic original conversation was repeated to mark the occasion with Bell in New York and Watson in San Francisco. But this time when Bell said, 'Mr Watson, come here, I want to see you', his exasperated former assistant had to point out that it would take him several days and a considerable amount of money to get there.

On his marriage Alexander Graham Bell had assigned all his shares in the Bell Telephone Company to his wife and father-in-law and he divided the remainder of his life between the US and Canada. The Bell Telephone Company would become the American Bell Telephone Company in 1885 and then the mighty A, T & T in 1899 for long the world's largest telecommunications company. The Alexander Graham Bell Laboratory that Bell founded in Washington in 1880 would become Bell Laboratories in New Jersey, a leading institution in

20th-century technological breakthroughs with seven Nobel Prizes for Physics awarded to various Bell Laboratories employees over the years as well as many technological breakthroughs including as far as the ghost of Alexander Bain was concerned the poignant sending of the first colour fax in 1924.

With Bell now a wealthy man he continued to dedicate himself to invention: he came up with an early form of the metal detector and was an early pioneer in aviation, but he will, of course, be forever associated with the telephone and his name is honoured in the decibel (dB), most often used as a measure of sound level. When Alexander Graham Bell died in 1922 more than fourteen million telephones in North America fell silent in his memory, a mark of respect that would have undoubtedly pleased him, for despite living abroad for more than fifty years he remained a true Scottish man at heart; he did have a telephone in his home but he refused to answer it, and instead stuffed it with paper.

DUE TO UNFORESEEN CIRCUMSTANCES THE POST OFFICE WILL BE CLOSED THIS EASTER

97 JAMES CONNOLLY (1868–1916)

The nation of Scotland would be fundamentally different without the influence of Ireland. The Scots take their name from the Irish people who from the 5th century travelled across the water and established their kingdom of Dalriada in Argyll which by the 10th century had taken over the rest of the country. Donegal-born monk Colm Chille travelled to Iona in the 6th century and as St Columba became the most famous figure in the history of Scottish Christianity. Belfast-born William Thomson (Lord Kelvin) moved to Glasgow in 1832 at the age of eight and became one of the most famous scientists in the world. Sligo-born Brother Walfrid founded Celtic Football Club in 1888, and when he left Glasgow for London in 1893 Walfrid donated the famous biscuit tin that has been used for the club's finances ever since.

And what has Scotland ever given Ireland in return? Well, other than 100,000 Protestant Lowlanders who emigrated to Ulster in the 17th century, a Scot called John Jameson bought a Dublin distillery in 1780 and began distilling Jameson's Irish Whiskey, the best selling Irish whiskey in the world – triple distilled from un-malted barley. The Jameson family motto is '*sine metu*', meaning 'without fear', and it appears on every bottle, although 'without ice' is also perfectly acceptable.

Another Scot who was *sine metu* was James Connolly. He was born in the Cowgate area of Edinburgh, the son of poor Irish immigrants, and at the age of fourteen he joined the British Army and served in Ireland – an experience that would change his life and turn him into a dedicated supporter of Irish independence. Returning to Scotland, Connolly became involved in socialist politics before moving to the US and then back to Ireland in 1910. Here, he became prominent in the Irish trade union movement and was one of the founders of the Irish Labour Party in 1912. By 1914 the long-promised bill for Home Rule for Ireland had finally been passed, only for World War I to put its implementation on hold. This suspension of Home Rule was reluctantly agreed to by the majority of Irish Nationalists; they had waited more than 300 years so another one or two could be just about endured, and Irishmen from both south and north joined in the fight for King and Country. For a Republican and a Socialist such as Connolly, however, neither fighting on behalf of the British nor the delay in Home Rule could be tolerated and he secretly, alongside other Republicans, planned what would become one of the most iconic events in modern Irish history.

On Easter Monday 1916 more than 1,000 Irish Republicans began their armed uprising and took control of buildings in central Dublin. Patrick Pearse was the main leader overall, but James Connolly was in charge of the Dublin men and it was he who captured the General Post Office building on O'Connell Street where the rebels set up their headquarters. The Easter Rising continued until the following Saturday when after British troops had been shelling Dublin's city centre for three days, the rebels surrendered. For all the aspirations of some rebel leaders there never had been any realistic likelihood of success; outside Dublin they had minimal support and by the time they surrendered the British outnumbered them by at least ten to one. Connolly, more than most, was under no illusion and had not expected the rising to achieve its ultimate goal of an Irish Republic, but a Marxist Socialist has to do what a Marxist Socialist has to do.

By the time the Rising was over, 130 military and police had been killed, 63 rebels and 250 civilians.

Connolly had correctly assumed that if they were unsuccessful then he and the other leaders and signatories of the Proclamation of the Irish Republic would probably pay for their rebellion with their lives, and the reprisals meted out by the British were, indeed, severe. Connolly had been seriously wounded in the fighting, but this did not prevent him along with fourteen others being sentenced to death by firing squad in early May 1916. When the execution order was carried out

Connolly, too ill to even stand up, was seated on a chair and shot and his body was dumped with the others in a mass grave.

The potency of a martyr is a powerful thing, where would Scottish nationalism be without the image of a hung, drawn and quartered William Wallace [1]? The merciless death of the mortally wounded Connolly was, more than anything else, the event that turned the Easter Rising from a misguided rebellion by a few extremists into the catalyst for a more fervent Irish Nationalism that would lead to the Irish War of Independence and the creation of the Irish Free State only five years later in 1921.

James Connolly became celebrated as one of the greatest Irish heroes, with Connolly Railway Station in Dublin named in his honour. But true socialist that he was, Connolly would have been less than impressed that the Irish Republic he gave his life for would under Eamon de Valera, one of the rebel survivors, become one of the most conservative nations in Western Europe. The GPO on O'Connell Street, that had been destroyed during the fighting, was finally rebuilt and reopened in 1929 and remains a symbol of the struggle for Irish Independence, although even today staff are still a little nervous when they return to work after the Easter holidays.

BUILD IT WITH ROSES

98 CHARLES RENNIE MACKINTOSH (1868–1928)

Scotland is famous for its landscapes and its architecture and the term, 'landscape architecture', was first coined by the art historian Gilbert Laing Meason of Forfar who in 1828 published a book called *On the Landscape Architecture of the great Painters of Italy*. Meason's term was then adopted by 19th-century botanist John Loudon, from Cambuslang in Lanarkshire, who decided that in the burgeoning field of garden design, landscape architecture and the occupation of landscape architect were rather more befitting than the alternative job title, intermittent planter of trees.

The genius of 18th-century architect Robert Adam is renowned. Adam was born in Kirkcaldy in 1728, the family moved to Edinburgh when he was aged eleven and there was little doubt about what young Robert was going to be when he grew up; his father, William, was a Scottish architect of considerable note and his brothers John and James were architects; only Uncle Fester had a different vocation. After travelling in Europe Robert Adam set up a practice in London in 1758 and soon became the pre-eminent and most fashionable architect in Britain, designing in a unique style that came to be known as neo-classicism. In Scotland, Adam's best-known buildings include Edinburgh University's Old College and Culzean Castle in Ayrshire, but in England his forte was grand, stately homes, which resulted in aristocrats across the country, including King George III, spending most of the second half of the 18th century having the builders in.

Between 1788–1822 Robert Adam had three volumes of his designs published called *The Works in Architecture*. This boosted Robert's international reputation considerably with the Adam style especially popular in the new republic of the United States where it was renamed Federal style and remained in vogue until 1820.

Robert Matthew Johnson Matthew from Edinburgh, RMJM for short or Rumjum for even shorter, is one of the world's largest architectural firms and RMJM (Scotland) were co-architects on the Scottish Parliament's Holyrood project and also gave Fred Goodwin a job when he was down to his last few million. In Scotland, the Parliament building was the subject of massive controversy and trenchant criticism, but it gained critical acclaim in the wider world where the general view was that the building would have been even nicer if they had just spent a bit more money on it.

The domestic opprobrium for the Holyrood building

would doubtless have come as no surprise to artist and architect Charles Rennie Mackintosh. Born in Glasgow, more than seventy years after Robert Adam's death, Mackintosh spent almost his entire curtailed architectural career in his native city. He was a designer within the British Arts and Crafts Movement headed by William Morris and, in addition, the main exponent of Art Nouveau in Britain. Mackintosh trained as an architect in Glasgow and studied drawing at the Glasgow School of Art where he met his future wife, English-born artist Margaret Macdonald, as well as her sister, the artist Frances Macdonald, and Frances' future husband, Glasgow-born architect Herbert McNair. This group became known as The Four, because there were four of them, and their work was significant in developing what became known as the Glasgow Style – the style for which Mackintosh is the undisputed figurehead. While the group's eclectic work did not find favour in Victorian Britain, it found high praise when they exhibited in Vienna in 1900.

In tandem with his collaboration with The Four, Mackintosh continued his work as an architect with the Glasgow firm Honeyman and Keppie from 1899–1913 and it was during this period that he designed the buildings for which he is most famous – Hill House in Helensburgh, completed in 1904, and Glasgow School of Art, completed in 1909. Compared to other architects of such high standing, Mackintosh's portfolio is relatively slender, but his influence on 20th-century building design was immense. Bold, straight lines, dramatic use of space and unified design schemes featuring striking but exceedingly uncomfortable high-backed chairs contributed to Mackintosh's unique style, but it was one which sadly proved far too modern for the Glasgow of his day.

Disillusioned by the lack of commissions, Charles Rennie Mackintosh and Margaret Macdonald left Glasgow for London in 1913, seven years later Mackintosh abandoned architecture altogether and in 1923 the couple moved to France where he turned to painting watercolours. When he died in London in December 1928 the work of Mackintosh had been all but forgotten in his native Scotland and the Glasgow tourist industry had to get by with day trips to Largs. Europe, however, was always receptive and appreciative of his work; for example, when the German architect Hermann Muthesius visited the Mackintosh-designed Willow Tearooms in Glasgow – and no doubt partook of proprietor Kate Cranston's delicious scones – he was inspired to help found the Deutscher Werkbund in 1907, a highly influential association of architects, artists and designers promoting modernity in design across the board, from industry to the home. Deutscher Werkbund would in turn inspire the sparse, geometric Bauhaus style, motivated by Walter Gropius, which sought to combine fine art, technology and functionality in designs for the modern age; at its height in the 1920s and early-30s Bauhaus exerted a virtually universal influence on subsequent developments in architecture, design, art and more.

Posthumously, Mackintosh was held up as an icon by architects and designers throughout the world, but it would not be until the city of Glasgow became the European City of Culture in 1990 and in search of some culture to celebrate that it unearthed Mackintosh, realised his impact elsewhere and went about celebrating his local works of genius. In the city of his birth, Mackintosh became ubiquitous; no postcard, fridge magnet or tea towel was complete without his famous rose design, or something vaguely resembling it. We will never know what Charles Rennie Mackintosh would have made of this belated popularisation of his work, but as he would have recognised more than most everyone has to make a living.

WHERE EXACTLY IS BHUTAN?

99 JOHN LOGIE BAIRD (1888–1946)

When the question of Scottish independence comes up several questions are always trotted out as reasons to keep the status quo: What will happen to the armed forces? Will there be checkpoints on the border? How can a small economy possibly manage to rack up the colossal national debt that seems to be so fashionable these days? And, most important of all, will we still get to see *Coronation Street*? Which only goes to show that television was perhaps the most influential invention of the 20th century and the reason why the SNP have consistently stated that they will ring-fence *Emmerdale* in any post-independence settlement.

John Logie Baird was born in the well-to-do Dunbartonshire town of Helensburgh, where no resident has ever been known to watch commercial television, and was so fascinated by the power of electricity that in his childhood he installed electric lighting at the family home and was able to get around the age-old problem of children being allowed access to the phone by having his own telephone exchange in his bedroom. Baird studied electrical engineering in Glasgow, but his progress was interrupted by chronic ill health that, throughout his life, caused him always to feel cold. This illness would lead to him leaving Scotland in 1923 for the slightly warmer climate in Sussex, but not before in one of his more ambitious electrical experiments he managed to black out most of Glasgow. Despite his poor health Baird was determined to make a living as an inventor, and after painfully failing to invent a new haemorrhoid treatment, he became obsessed with the concept of 'television', or 'seeing by wireless', which had been the holy grail for electrical engineers and scientists for decades.

In January 1924 in his lodgings in Hastings, Sussex, Baird, by connecting a Nipkow disk – a spinning cardboard disk punctured by a series of holes in a spiral pattern – to an assortment of lenses, electric motors, torch batteries, household utensils all glued together, found that he was able to transmit the image of a Maltese cross over a distance of ten feet (3 m). However the potential of his new technology was almost very short-lived indeed when in July 1924 Baird received a 1,200-volt electric shock that threw him across his workshop. Incredibly, Baird survived with nothing more than burned hands and a Jedward hairstyle, but it was the final straw for his landlord who demanded that the inventor leave forthwith, never mind Baird's promise that he would get him a free Sky Sports subscription.

Baird moved to London where, after further demonstrations of silhouettes in Selfridges department store, by October 1925 in his new Soho workshop he succeeded in transmitting the first live, moving, television images; tones of shade and light were discernible and the subject in his image was a ventriloquist's dummy called Stooky Bill. Baird demonstrated his achievement to invited members of the Royal Institution on 26 January 1926, using a television system comprising thirty lines per resolution and five frames per second and with a human face in place of the ventriloquist's dummy. Sadly, it cannot be confirmed whose face it was that made the first ever appearance on television, but speculation suggests that it could well have been a fresh-faced Dougie Donnelly.

It was just as well that Baird was able to demonstrate his mechanical television when he did, as not only had he spent all his Selfridges gift vouchers, but also other inventors were developing television, including Americans Charles Jenkins, Philo Farnsworth and Vladimir Zworykin, and in the US the first long-distance of mechanical television took place in 1927. The demonstration was organised by Herbert E. Ives of the Bell Company, whose founder was the by then late Alexander Graham Bell [96], and images were transmitted from Washington to New York. In response, John Logie Baird immediately set up the Baird Television Development Company and in 1928 was the first to demonstrate transmission across the Atlantic. In September 1929 he began his first experimental broadcasts on the BBC network, and in 1930 the first simultaneous broadcast in sound as well as vision.

The BBC broadcast was not, however, the world's first. Experimental broadcasts had started in New York in 1928, and in the same year American Philo T. Farnsworth had demonstrated the first fully electronic television system. Backed by the big American communications companies, electronic systems began to outstrip Baird's mechanical system no matter how many improvements he made. By 1935 Baird's broadcasts for the BBC had increased to a resolution of 240 lines, but in 1937 the Beeb, which under Director-General John Reith [84], had never been especially welcoming to either television or Baird, ditched him and his television system for a 400-line American alternative. Disappointed but undeterred, Baird continued to do what he always did and sought to improve upon his system. He had demonstrated the world's first colour television transmission in 1928, the world's first colour television broadcast in 1939 and in 1944, two years before his death, demonstrated a 600-line high-definition electronic colour television system that has since proved to be sixty years ahead of its time.

John Logie Baird who had bravely fought ill-health for most of his life died in 1946 when television was still in its infancy, but being the pioneer and genius that he was Baird had predicted correctly that television would become the world's dominant form of communication, as has been proved beyond doubt in the second half of the 20th century. The final country to succumb to the power of television was the Asian kingdom of Bhutan where the government, whose policies have often been portrayed as isolationist, had enforced a long-standing ban. But concluding, presumably, that they could not avoid this particular outside influence forever, the first programmes ever to be shown in Bhutan were screened in 1999 and, honouring of the nationality of the man who invented television, began with an episode of Scottish entertainment show *The Hour*, an experience so disturbing that most Bhutanese resolved never to watch television again.

NO FRIENDLY BOMBS FALLING ON SLOUGH, I'M AFRAID

100 ROBERT WATSON-WATT (1892–1973)

James Clerk Maxwell [76] was so far ahead of his scientific contemporaries that when he wrote in 1865 about the existence of radio waves it would be more than twenty years before German scientist Heinrich Hertz actually discovered them in 1887. And it was another German, Christian Huelsmeyer, in 1904 who was the first to publicly demonstrate using radio echoes to detect a ship's presence, albeit within a range of only two miles (3 km). However, as Huelsmeyer's invention could not pinpoint the ship's position, it was concluded that it was better in conditions of poor visibility for maritime collisions to be an unexpected surprise rather than spending all your time in a blind panic about where all the other boats might be.

It would take another thirty years for radar as we know it today, meaning radar detection and ranging, to become a practical reality. During those thirty years, so many countries, different engineers and scientists have played a part in the development of radar technology that it is almost impossible to name any one individual as the inventor. Having said that, if the title did have to be awarded to one person, then it would go to a meteorologist from the historic Angus town of Brechin, famous for its cathedral and Flicks nightclub and where one can only speculate about what manner of mishaps take place when the north-east Haar creeps in.

Robert Watson-Watt, a descendant of James Watt's [52], graduated from Dundee University with a degree in engineering before moving to Farnborough in England where he worked as a meteorologist from 1915. Watson-Watt had become fascinated by radio waves at university and in his new job began researching whether radio, by picking up lightning flashes, could detect and locate thunderstorms, thus serving as a weather warning for aircraft pilots. By 1923 he had succeeded and in the process put in place the science underlying radar, technology he would continue to develop throughout the 1920s. In 1927 he was promoted to become head of the Radio Research Station in Slough before becoming Superintendent in 1933 of the Radio Department of the National Physics Laboratory.

At the same time, scientists and engineers in the US, Germany and the Soviet Union were all, also, working to develop and apply methods of radio detection, and all the more imperative when Hitler came to power in 1933 and began to rearm. Governments throughout the rest of Europe were all too aware of the considerable strength of Hitler's air force and were desperate to put in place any early warning systems they could. In 1934 the Americans and the Russians had discovered that, at short range, radio waves bounced off airplanes and ships and in the same year the British set up a committee under Henry Tizard to urgently investigate this further.

Watson-Watt was asked if he could invent some sort of 'death-ray' that would blow German planes out of the sky but, sadly, the technology behind such a weapon could be found only on Planet Mongo and Ming the Merciless was not willing to share. Instead, in February 1935 Watson-Watt and his assistant, Arnold Wilkins, drafted a report entitled, *The Detection of Aircraft by*

Radio Methods and presented it to Henry Tizard. On 26 February 1935, at Daventry radio station, the first practical demonstration of high-frequency radar for the purpose of aircraft detection took place: two antennae, by process of receiving the radio waves deflected from the plane, successfully detected a bomber at a range of up to eight miles (13 km). More progress was made and by March the following year the range had increased from eight miles (13 km) to ninety miles (145 km), far ahead of anything being achieved by other countries.

Robert Watson-Watt was granted a UK patent in April 1935, put in charge of a team to work with and set up in a top-secret research station in Suffolk. By 1937, with Hugh Dowding [19] who was in charge of RAF Fighter Command an enthusiastic supporter, the team had begun to build the first early-warning radar towers designed to detect aircraft at great distance and in all weathers, and within two years nineteen of these Chain Home towers with a detection range of 120 miles (192 km) were in place along the English coastline.

At the outbreak of World War II all the major powers had developed radar programmes but, thanks to Watson-Watt and Arnold Wilkins alongside their fellow scientists, the RAF, politicians, indeed everybody who had worked toward putting an early warning system in place at the first opportunity, Britain gained an advantage that proved crucial to victory over the Luftwaffe in the Battle of Britain in 1940: British pilots knew in advance where the enemy was.

During the war Watson-Watt worked as an advisor to the Air Ministry and when it was all over, he went to live first in Canada and then in the US before returning to Britain. By then, radar was in wide, routine use around the world in, among other areas, air traffic control, as a navigational aid and for catching speeding motorists. One speeding driver caught out with a radar gun was none other than Robert Watson-Watt. When he was flagged down in Canada in 1956 and fined $12 Watson-Watt is reported to have said, 'Had I known what you were going to do with it I would never have invented it'.

Epilogue: Wha's Like Us #2.0

Reflecting on my one hundred Great Scots, it is clear that there is no area of life, and few places on earth, upon which the nation's people have not brought influence to bear: influence and innovation that has, without doubt, changed the world. And if history is an honest indicator of their future then Scots will remain at the forefront in shaping lives and the world we live in through the 21st century and forever more.

I end my discourse as I began with reference to T. Anderson Cairns's prose *Wha's Like Us*. Through the story of a day in the life of an Englishman Anderson Cairns points up the myriad inventions, all by Scots, which the man comes into contact with from the moment he awakes.

The inventors name-checked are Charles Macintosh, John Loudon McAdam, John Boyd Dunlop, James Chalmers, Alexander Graham Bell, Kirkpatrick Macmillan, John Logie Baird, John Paul Jones, James VI, Patrick Ferguson, Alexander Fleming, James Young Simpson and William Paterson. However, such has been Scottish influence over the world that it seems only right to update Anderson Cairns's work for the 21st century and thus I give you *Wha's Like Us #2.0* – with due thanks to John Napier who popularised the use of the decimal point.

WHA'S LIKE US #2.0

Mr and Mrs Smith awake in the morning and check the time as standardised around the world by Sandford Fleming from Fife to make sure that they have not slept in. They then, one after the other, proceed to the bathroom for their morning ablutions, remembering to flush their toilet as patented by Alexander Cummings from Edinburgh.

They meet in the kitchen where they enjoy a breakfast of Lipton's Tea, named after Thomas Lipton from Glasgow, along with toast spread with marmalade as created by Janet Keiller from Dundee, and each has a glass of vitamin C-rich fruit juice as recommended by James Lind from Edinburgh who discovered the health benefits of citrus fruit; the fruit juice has been nicely chilled through artificial refrigeration as first demonstrated by William Cullen from Hamilton. As they eat their breakfast they might turn on the radio, with the existence of radio waves predicted by James Clerk Maxwell from Edinburgh, and catch up with the news on BBC as founded by John Reith from Stonehaven, paying special attention to current events in the financial markets that look to Adam Smith from Kirkcaldy as the Father of Modern Economics.

In his office, Mr Smith notices that a message has come through on his fax machine, as invented by Alexander Bain from Caithness. The information on the fax is somewhat worrying but, luckily, to alleviate his stress he has to hand some beta-blockers, as invented by James Black from Uddingston. Mrs Smith has some errands to run and goes to the local shop where she has to buy some 60-watt light bulbs, with the watt named after James Watt from Greenock and the light bulbs possibly invented by James Bowman Lindsay from Angus, and then she heads off to the Royal Bank of Scotland, hoping that the bank has not gone bust again, to withdraw some money from the ATM, as invented by James Goodfellow from Paisley.

After his stressful morning, Mr Smith contemplates a round of golf, as first played in Scotland, but decides that the garden is more important and gets out the lawnmower with its two-stroke engine invented by Dugald Clerk from

Glasgow. Mrs Smith goes to the local pool for a swim and will try out her new Speedo swimwear as invented by Alexander McRae from the Kyle of Lochalsh.

With the lawn successfully mowed and Mrs Smith back from the pool it's time for a well-deserved drink and she opts for a refreshingly crisp Australian shiraz from Hunter Valley, named after John Hunter from Leith, and not forgetting James Busby from Edinburgh who was the first to plant vines there.

At the end of the day, while listening to the sweet sound of AC/DC from Glasgow via Melbourne, Mr and Mrs Smith look from their window toward the nearby wind farm, with the first electricity-producing wind turbine invented by James Blyth from Angus, and ponder, as David Hume from Edinburgh might have, about their place in the universe and reflect how their lives are but a speck of time on the earth that James Hutton from Edinburgh first proved to be truly millions of years old. And they muse over the great questions yet to be answered such as: Will Andy Murray ever win a Grand Slam? Will Alex Ferguson ever retire? And whatever happened to that grumpy Scottish bloke with the nice wife who used to run the country?

Wha's like us – other than the tens of millions of Americans, Canadians, Australians, New Zealanders and Northern Irish.

Damn few, and as the great Scottish Presbyterian John Knox with his faithful collie Shep would say, they are all going to Hell anyway.

More Non-fiction from Headline

Caledonia

JOHN [illegible]

[illegible]

Caledonia [illegible]

More Non-fiction from Hachette Scotland

Caledonication

JOHN K.V. EUNSON

The story of Scotland is dramatic, fascinating, at times tragic and generally quite wet and windy.

In this entertaining and mostly accurate retelling of a history of a nation, we relive such momentous events as Bannockburn, the Reformation and Argentina '78. We re-examine the discovery of famous Scottish inventions including the steam engine, the telephone, penicillin and Irn-Bru. And the analysis of the extraordinary influence of Scots around the world from Pontius Pilate to Groundskeeper Willie will inform and amaze.

Caledonication is a completely up-to-date and easily transportable account of everything you need to know about the land that is, to some, 'the best small country in the world' and, to others, somewhere in England.

NON-FICTION / HUMOUR 978 0 7553 1858 2